Photovoltaic Modules

Also of interest

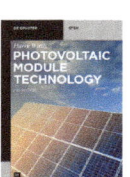

Photovoltaic Module Technology
Harry Wirth, 2021
ISBN 978-3-11-067697-6, e-ISBN (PDF) 978-3-11-067701-0
e-ISBN (EPUB) 978-3-11-067710-2

Energy and Sustainable Development
Quinta Nwanosike Warren, 2021
ISBN 978-1-5015-1973-4, e-ISBN (PDF) 978-1-5015-1977-2
e-ISBN (EPUB) 978-1-5015-1328-2

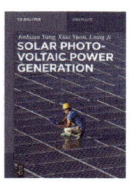

Solar Photovoltaic Power Generation
Jinhuan Yang, Xiao Yuan and Liang Ji, 2020
ISBN 978-3-11-053138-1, e-ISBN (PDF) 978-3-11-052483-3
e-ISBN (EPUB) 978-3-11-052542-7

Electrochemical Energy Storage.
Physics and Chemistry of Batteries
Reinhart Job, 2020
ISBN 978-3-11-048437-3, e-ISBN (PDF) 978-3-11-048442-7
e-ISBN (EPUB) 978-3-11-048454-0

Applied Electrochemistry
Krystyna Jackowska, Paweł Krysiński, 2020
ISBN 978-3-11-060077-3, e-ISBN (PDF) 978-3-11-060083-4
e-ISBN (EPUB) 978-3-11-059897-1

Photovoltaic Modules

Reliability and Sustainability

Edited by
Karl-Anders Weiß

DE GRUYTER

Editor
Dr. Karl-Anders Weiß
Fraunhofer Institute for Solar Energy Systems ISE
Energy Technologies and Systems
Freiburg, Germany
karl-anders.weiss@ise.fraunhofer.de

ISBN 978-3-11-068554-1
e-ISBN (PDF) 978-3-11-068555-8
e-ISBN (EPUB) 978-3-11-068572-5

Library of Congress Control Number: 2021933395

Bibliographic information published by the Deutsche Nationalbibliothek
The Deutsche Nationalbibliothek lists this publication in the Deutsche Nationalbibliografie;
detailed bibliographic data are available on the Internet at http://dnb.dnb.de.

© 2021 Walter de Gruyter GmbH, Berlin/Boston
Cover image: Karl-Anders Weiß, idea Elisabeth Klimm
Typesetting: Integra Software Services Pvt. Ltd.
Printing and binding: CPI books GmbH, Leck

www.degruyter.com

MIX
Papier aus verantwor-
tungsvollen Quellen
FSC® C0834 1

Preface

In September 2015, the United Nations adopted its Sustainable Development Goals,[1] the blueprint to achieve a better and more sustainable future for all. The decision was directly followed by the 2015 Climate Change Conference in Paris sending a clear signal for a worldwide energy transition towards renewables, a transition to take place long before the total consumption of fossil and nuclear resources and timely before catastrophic climate change.

Photovoltaic (PV) electricity generation is developing into a major global player in our future energy supply scenarios, enabled by a cost-reduction momentum not foreseen 20 years ago even by most optimistic PV supporters. Recent studies find generation costs for PV electricity in the range of 2–6 ct(€)/kWh, depending on the size and location of the power plant. The cost learning curve will keep progressing since many innovations are well underway in all parts of the value chain. PV modules are the key components for every PV power plant from tiny roof-top systems of a few kilowatts to plants in the gigawatt range which may demand millions of modules. Modules are required to efficiently, safely, and sustainably convert solar irradiance into electric power over a service life of decades. For such long operational times, reliability is the dominating property of a product.

The first part of this book addresses crystalline silicon, wafer-based module reliability. It gives an overview on the very important topic of module reliability which is crucial for the long-term operation and also for the investments in PV systems. It describes analytical methods for module and material characterizations, relevant loads for PV modules, and the design of accelerated aging tests adapted to PV module technology. This is followed by a description of how reliability tests for materials and modules can be developed. The relevant international standards for type approval and safety testing of modules and materials are described, as well as their meanings and limitations. In the end, methods for degradation modeling and service life prediction are introduced.

The second part of the book addresses the very important topic of sustainability of PV modules, which is especially crucial for renewable energy products. An overview on sustainability assessment methods is given as well as insights on the influences of the different components of PV systems on the sustainability of PV electricity. Methods to determine the ecological inventory of the different steps of the value creation chain, from basic materials to recycling, are presented. The part ends with a short overview on certification and legislation measures addressing sustainability.

---

1 www.un.org/sustainabledevelopment/sustainable-development-goals/

https://doi.org/10.1515/9783110685558-202

The authors would like to thank their colleagues in different institutions for many valuable discussions and generous contributions to this book from their various R&D projects. Karl-Anders Weiß would like to thank all coauthors and proofreaders for their willingness to support the new edition of the book and the longtime cooperation and friendship this project is built on.

Contents

Part I: Crystalline silicon module reliability

Part II: **Crystalline silicon module sustainability**

Symbols and units

α	1/m	Absorption coefficient
α	°	Angle
A	m^2	Area, cross section
A	1	Absorptance
β	°	Angle
d	m	Diameter, distance, layer thickness
E	W/m^2	Irradiance
E	J	(Electric) energy
E_λ	W/m^3	Spectral irradiance
FF	1	Fill factor
H	MJ/m^2 (kWh/m^2)	Irradiation (1 kWh = 3.6 MJ)
I	A	Current
$K_{\tau\alpha}$	1	Incidence angle modifier
L	m	Edge length
AM	1	Air mass
n	1	Refractive index or counting index
P	W	Power
PR	1	Performance ratio (of PV module or PV power plant)
QE	1	Quantum efficiency
R	1	Reflectance
r	Ω	Sheet resistivity
ρ	$\Omega \cdot$ m	Electrical (volume) resistivity
T	1	Transmittance
T	°C, K	Temperature in degree Celsius or in Kelvin as indicated
V	V	Voltage
v	m/s	Velocity
w	m	Width

Subscripts

a	activation
AC	Alternating current
beam	Beam or direct (irradiance, irradiation)
Bulk	Referring to the material volume, without interfaces
cell	Referring to the cell
DC	Direct current
diff	Diffuse (irradiance, irradiation)
e	Electron
eff	Effective
enc	Encapsulant
ext	External
f	fitted
glob	Global (irradiance, irradiation)
h	Hydrolisis
int	Internal

https://doi.org/10.1515/9783110685558-204

m	measured
max	maximum
mod	PV module
MPP	Maximum power point
OC	Open circuit
ox	oxidized
P	Production
PD	Photodegradation
plant	PV power plant
POA	Plane of the array
red	reduced
ref	reference
res	residual
SC	Short circuit or solar constant
STC	Standard testing conditions
T	Thermal
Tm	Thermomechanical
TIR	Total internal reflectance
U	Usage

Abbreviations

AR	Antireflective
AES	Auger electron spectroscopy
AFM	Atomic force microscopy
AM	Air mass
AS	Antisoiling
ASTM	American Society for Testing and Materials
B2B	Business to business
B2C	Business to customer
BOM	Bill of materials
BOS	Balance of System
BST	Black standard temperature
CB	Certification body
CdTe	Cadmium telluride
CED	Cumulated energy demand
CIE	Commission Internationale de l'Ecleirage
CIS	Copper indium selenide
c-Si	Crystalline silicon
CPV	Concentrating PV
CTE	Coefficient of thermal expansion
CTM	Cell to module
DH	Damp heat
DIN	Deutsches Institut für Normung – German Standardization Organization
DKE	Deutsche Kommission Elektrotechnik – German Electrotechnical Commission
DMA	Dynamic mechanical analysis
DoC	Degree of cross-linking

DSC	Differential scanning calorimetry
DUT	Device under test
EDX	Energy-dispersive x-ray spectroscopy
EL	Electroluminescence
EN	European standard
EoL	End of life
EPBT	Energy payback time
EQE	External quantum efficiency
EU	European Union
EVA	Ethylene-vinyl acetate
FIT	Feed in tariff
FL	Fluorescence
FT-IR	Fourier-transformation infrared spectroscopy
FRBG	Freiburg, Germany
FU	Functional unit
G	Global irradiation
GC	Gran Canaria
GEC	US Green Electronics Council
GIS	Geographical information system
GPP	Green Public Procurement
HF	Humidity freeze
IAM	Incidence angle modifier
IEC	International Electrotechnical Commission
InGaAs	Indium gallium arsenide
ILCD	International Reference Life Cycle Data System
IQE	Internal quantum efficiency
IR	Infrared
ISO	International Standardization Organization
LCA	Life cycle assessment
LCIA	Life cycle impact assessment
LVD	European low-voltage directive
MDG	Millennium Development Goal
ML	Mechanical load
mono-Si	Monocrystalline silicon
MPP	Maximum power point
MQT	Module quality test
MST	Module safety test
NEC	National Electric Code
NEG	Negev desert, Israel
NMOT	Nominal module operating temperature
NOCT	Nominal operating cell temperature
OTR	Oxygen transmission rate
PEF	Product environmental footprint
PA	Polyamide
PE	Polyethylene
PET	Polyethylene terephthalate
PID	Potential-induced degradation
PL	Photoluminescence
POA	Plane of array

POE	Polyolefin elastomers
poly-Si	Poly- or multicrystalline silicon
PP	Polypropylene
PV	Photovoltaic(s)
PVB	Polyvinyl butyral
PVDF	Polyvinylidene fluoride
PVF	Polyvinyl fluoride
QA	Quality assurance
RH	Relative humidity
SAM	Scanning acoustic microscopy
SD	Standard deviation
SDG	Sustainable developments goal
STC	Standard testing conditions
TC	Temperature cycling
TCO	Total cost of ownership
TF	Thin film
TMF	Thermomechanical fatigue
TOW	Time of wetness
TPSE	Thermoplastic silicone elastomer
TPT	Backsheet laminate with layers Tedlar-PET-Tedlar
UFS	Environmental Research Station on the mountain Zugspitze, Germany
UN	United Nations
UV	Ultraviolet
vis	Visual
WEEE	Waste electrical and electronic equipment directive
WST	White standard temperature
WVD	Water vapor diffusion
WVTR	Water vapor transmission rate
YI	Yellowness Index

About the editor

Karl-Anders Weiß, born in 1979 in Schwäbisch Hall, Germany, studied Physics and Economics at the University of Ulm and holds a Ph.D. in Physics. He joined the Fraunhofer Institute for Solar Energy Systems ISE in 2005 and focuses on reliability and service life prediction of materials and components. As these are strongly linked to aspects of sustainability, this became a second focal topic of his work in recent years. He has supervised numerous B.Sc., M.Sc., and Ph.D. students and is member of several scientific and standardization committees related to environmental engineering and sustainability.

https://doi.org/10.1515/9783110685558-205

List of contributing authors

Dr. Karl-Anders Weiß
Fraunhofer Institute for
Solar Energy Systems ISE
Energy Technologies and Systems
Freiburg, Germany
karl-anders.weiss@ise.fraunhofer.de

Sina Herceg
Fraunhofer Institute for
Solar Energy Systems ISE
Service Life Analysis and Material
Characterization
Freiburg, Germany
sina.herceg@ise.fraunhofer.de

Dr. Bengt Jäckel
Fraunhofer Center for Silicon
Photovoltaics (CSP)
Module and System Reliability
Halle, Germany
bengt.jaeckel@csp.fraunhofer.de

Dr. Ismail Kaaya
Fraunhofer Institute for
Solar Energy Systems ISE
Service Life Analysis and Material
Characterization
Freiburg, Germany
ismail.kaaya@ise.fraunhofer.de

Elisabeth Klimm
EK Reliability
Wittnau, Germany
e.klimm@ek-reliability.de

Dr. Gernot Oreški
Polymer Competence Center Leoben GmbH
Smart Material and Surface Testing
Leoben, Austria
gernot.oreski@pccl.at

Sebastián Pinto Bautista
Helmholtz-Institut HIU Elektrochemische
Energiespeicherung
Technology Assessment and System Analysis
Karlsruhe, Germany
sebastian.bautista@kit.edu

Djamel Eddine Mansour
Fraunhofer Institute for Solar Energy
Systems (ISE)
Freiburg, Germany
djamel.eddine.mansour@ise.fraunhofer.de

https://doi.org/10.1515/9783110685558-206

Karl-Anders Weiß

1 Introduction

Photovoltaic (PV) modules directly convert solar radiation into electrical energy at appropriate voltage and current levels. The discovery of the photoelectric effect is credited to the French scientist Alexandre Edmond Becquerel. He built an electrolytic cell in 1839 that was able to deliver electric power when exposed to light. More than a century later, in 1954, Daryl Chapin, Calvin Fuller, and Gerald Pearson developed a silicon-based PV cell at Bell Laboratories, reporting efficiencies of up to 6% [1]. In the 1960s, the potential of PV modules to supply energy in remote applications attracted growing interest, for example, for lighthouses or offshore navigation signals, where utility grid connection or alternate energy sources were more expensive. The oil crisis of 1973 triggered research and development in alternative energy resources worldwide and also enhanced the support for research in PV. In 1975, the US government started a series of procurement activities that accelerated technology and testing development for PV modules. It also led to the development of adapted tests and further also to the development of type approval standards in the frame of the International Electrotechnical Commission. At the very beginning, encapsulated modules with glass covering appeared on the market, followed by the first laminated modules using polyvinyl butyral (PVB) as an encapsulant and a polyester film as a rear cover. The latter design, using the front cover as a rigid support, became and still is the predominant module design. In the late 1970s, Tedlar was introduced as a backsheet material. In the early 1980s, ethylene-vinyl acetate (EVA) became the predominant encapsulation material. Since then the basic module design has not been changed.

In the 1990s, Germany launched a substantial PV program to implement 1,000 roof systems ("The 1,000 Roofs Programme"), to be followed by a 100,000 roof program in 1999. These programs contributed to make grid-connected PV a recognized and important candidate for renewable electricity supply of reasonable scale. Japan and California also announced PV support schemes in the 1990s.

The German feed-in-tariff scheme issued in 2001 boosted PV mass production and led to tremendous cost reductions and market growth (Fig. 1.1). In the meantime, PV became more and more competitive compared to other electricity generation technologies and reached grid parity in many regions. It is one of the important sources for electricity supply now, and is expected to be even more in the future.

In terms of 2019 market share and installations, *wafer-based* crystalline silicon (c-Si) accounts for almost 95% of the produced PV capacity. Looking at the installed capacity, c-Si accounts for more than 90% of the total PV installations worldwide. The remaining share is mostly thin-film PV, which comprises mainly cadmium telluride (CdTe) and copper indium selenide (CIS) technologies. Market developments, especially the extreme reduction of prices for c-Si modules have to be mentioned, since

https://doi.org/10.1515/9783110685558-001

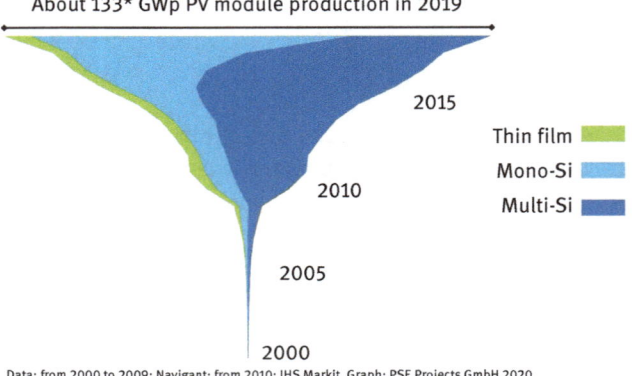

About 133* GWp PV module production in 2019

Data: from 2000 to 2009: Navigant; from 2010: IHS Markit. Graph: PSE Projects GmbH 2020

Fig. 1.1: Development of the production of PV modules since 2000, including monocrystalline, multicrystalline, and thin-film technologies. Source: PSE Projects GmbH.

this was the main driver for the reduction of PV electricity generation costs, having increased the relative market share of c-Si modules in recent years (Fig. 1.2).

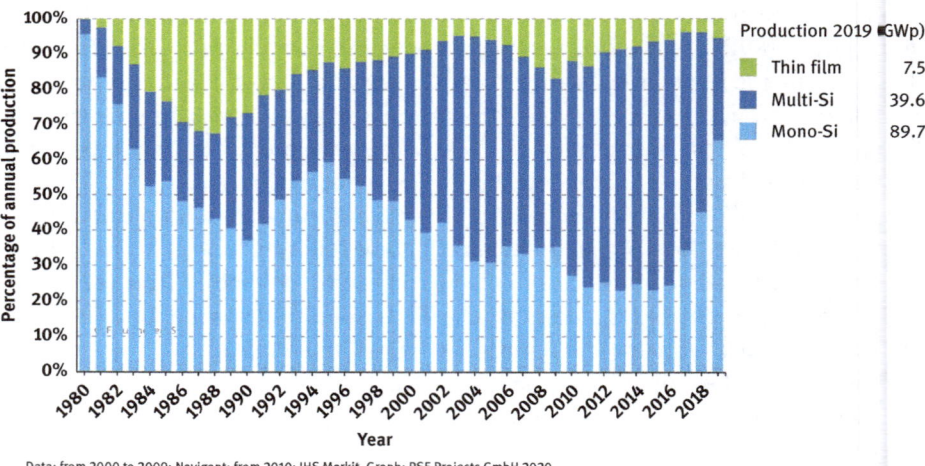

Data: from 2000 to 2009: Navigant; from 2010: IHS Markit. Graph: PSE Projects GmbH 2020

Fig. 1.2: Development of market share of PV module technologies since 1980, including monocrystalline, multicrystalline, and thin-film technologies. Source: PSE Projects GmbH.

The first part of this book deals with *wafer-based* c-Si module reliability. Since growth of the market can be seen in almost every country and region worldwide, additional questions related to reliability due to specific local requirements pertaining to climatic or operational loads have to be addressed. We give a general overview on loads affecting the reliability of PV modules, including characterization methods to measure effects. Further chapters give insights into testing of materials and modules as well as certification and service life prediction methodologies.

1.1 Module technology

PV modules contain a number of components and materials (Fig. 1.3), and specific wording is used to describe these parts. Here, a short explanation and definition is given to avoid misunderstandings related to the content of the book. More detailed descriptions of PV module technologies can be found in the literature, for example, in "Photovoltaic Module Technology" by Harry Wirth.

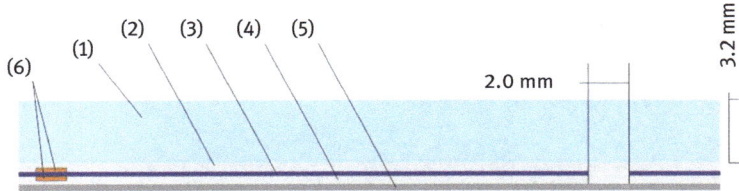

Fig. 1.3: Cross section of the laminate of a typical PV module. The marked components are described further. Source: Harry Wirth.

(1) Glazing: Front part of a PV module is transparent to lead incident sunlight to PV cells. Usually made from low-iron solar glass, there are also modules with polymeric glazing. Glazing usually also gives dimensional stability to the module. Typical thickness is 3–4 mm, in glass–glass modules sometimes <3 mm.

(2) and (4) Encapsulation: Transparent polymeric material connects/glues all components of the module laminate. Predominant material is EVA copolymer but there are also other materials used such as ionomers, PVB, polyolefins, or silicones. Thickness is typically 400–500 µm in front of cells and 400–500 µm between cells and backsheet.

(3) PV cells: Active semiconductor component in module generating electricity from incident sunlight. It can be made of different semiconductor materials, but here we focus on cells made from multicrystalline or monocrystalline silicon. The size typically varies between 156 × 156 mm^2 and 210 × 210 mm^2 (6″ cells) but can differ, and the trend goes to larger cells. Cell thickness is typically of 150–180 µm.

(5) Backsheet: It is the material at the rear side of the module providing electrical insulation and protection against climatic impact; typically made of polymeric materials, it is often produced as laminated or coextruded multilayer films. Used materials are polyethylene terephthalate (PET), polyamide (PA), polyvinyl fluoride (PVF), polyvinylidene fluoride (PVDF), polyolefins (PO), or other polymeric materials. The films usually have a thickness of 250–350 µm. Glass is also used as a material for the rear side of PV modules in glass–glass module types (for materials, see Glazing).

(6) Interconnectors: The metallic wires are used to electrically connect the cells and lead the generated electricity out of the modules. Typically, cells are connected by two to eight wires; there are also connection technologies with even more wires (multiwire) with different diameters of wires, depending on the number of wires. Wires are usually made of copper with tin coating. Thicker wires connecting several strings and carrying higher currents are usually called bus bars. The electrical connection is also shown in Fig. 1.4.

Fig. 1.4: Schematic of the electrical connection of a PV module and cross section showing the connection of module laminate and frame. Source: Harry Wirth.

Frame: Especially glass–backsheet modules usually contain a frame of aluminum profiles to enable mounting. Glass–glass modules are usually produced without an additional frame. The module laminate (components (1)–(6) in Fig. 1.3) are glued in the frame usually using silicone or adhesive tape. The basic design is shown in Fig. 1.4.

Junction box: The junction box is mostly glued to the rear side of the module or at the edge of the module at the position where the bus bars exit the module laminate. In Fig. 1.4, this position can be seen in the lower part of the picture where the bus bars meet in the black area. In the junction box, the bus bars are connected to the cables. The junction box is typically made of polymeric materials, and it provides electrical insulation for the live parts.

1.2 Terminology

To ensure that the descriptions and explanations in the book are easy and clear to understand, the major terms are defined further, as they are used throughout the book.

Reliability: Reliability is the likelihood that a product will not fail within a specific time period. This is a key element for users/owners/stakeholders that need the product to work without fail.

Service life: Service life is the total time of operation of a product. It can end due to technical reasons (e.g., failure and degradation) or due to economic decisions (e.g., replacement by a more efficient product, high costs for maintenance, and end of contract).

Sustainability: Sustainability describes the avoidance of the depletion of natural resources by a product in order to maintain an ecological balance. Usually, it is given as a comparison of different products or scenarios and rarer as absolute values. It contains a lot of parameters describing the impact of a product on the environment.

Climatic data and climatic loads are the major influence for degradation processes and thus also influencing reliability. There is typically a correlation between the ambient climatic conditions (the weather) and the specific loads for the sample. To make it easier to discuss climate-related loads, the following terms are used with the described meaning in this book:

Macroclimate: Macroclimate shows the climatic conditions in the surrounding of the sample, which is also called ambient climate. This includes the typically measured weather data like irradiation, temperature, humidity, precipitation, wind speed, and direction in outdoor operation.

Microclimate: Microclimate is the climatic load at a very specific position, for example, in or on a piece of material. These loads can differ clearly from the macroclimate even if they are correlated. For example, the temperature of a sample under irradiation is usually higher than the ambient temperature; humidity on the infinitesimally small layer of air close to the surface of the sample is also lower than in the ambient conditions. For the estimation and calculation of (material) degradation processes, the microclimate describes the relevant load conditions. This is of special importance if natural (outdoor) and accelerated aging effects are compared.

Part I: **Crystalline silicon module reliability**

Karl-Anders Weiß

2 Market-related topics of reliability

Photovoltaic (PV) systems became a major source for electricity supply on the global scale as the impressive graphs in Chapter 1 show, and in many areas, PV is no longer supported by subsidies or other legal measures since it can economically compete with other (classically fossil) sources even without subsidies. The basic difference of renewable energy systems – from the economic point of view – to fossil-driven power plants is that the share of operational costs is almost negligible for the renewable systems since they do not "use" any kind of fuel that has to be purchased. Therefore, the cost of the produced electricity is mainly influenced by the initial investment, and the operational costs are of much lower influence than for classical power plants. Over the lifetime of a power plant, the cost of the produced electricity is calculated as follows:

$$\text{cost(kWh)} = \frac{\int_0^{t_{lt}} P(t)dt}{\int_0^{t_{lt}} \text{Cost}(t)dt} \tag{2.1}$$

so the electricity costs are simply defined by the total yield of the power plant over the lifetime t_{lt} divided by the total costs over the lifetime. Thus, to reduce electricity costs, the following basic possibilities are available:

- reduction of costs (e.g., by optimizing purchasing or financial measures);
- increased power P_0;
- reduction of degradation dP/dt; and
- increased lifetime t_{lt}.

Two of these variables are directly linked to reliability, and this clearly shows the relevance of reliability for economic decisions. To optimize the economy of a PV system, which can be directly transferred to a PV module, there is a triangle system linking conflicting goals (Fig. 2.1). When one thinks of increased performance and improved durability, naturally higher costs are assumed, whereas reduced costs together with high initial performance lead to reduced durability. So, in short, the three parameters cannot be optimized individually without influencing the others. An integral optimization is required. For such optimization effects and processes, the influencing reliability and durability have to be known and understood, and reasonable testing is required.

The relevance of reliability and service lifetime of PV modules is still of special importance even when economic decisions are tight. Investments in PV systems are usually calculated for long term, typically more than 10 or 20 years. Even contracts with duration of 50 years have been reported. With these long contract durations, degradation effects have very strong influence. If economic conditions are changing while an existing system is in operation, even financial go or no-go decisions can

https://doi.org/10.1515/9783110685558-002

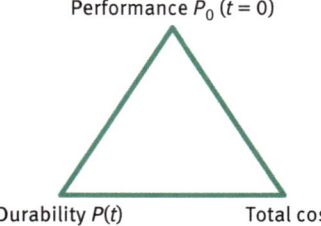
Performance P_0 $(t = 0)$

Durability $P(t)$ Total cost

Fig. 2.1: Triangle of influencing factors for economy of PV electricity: performance, durability, and cost.

depend on the expected degradation and lifetime. An important example for this effect appeared when the feed-in tariffs for PV systems were cut in Spain, also for existing plants. The financial viability of a lot of plants depended, due to the decision of the government in Spain to reduce the payment for the electricity, on an expectable lifetime of 25 years. Previously, with the initially planned payment per kilowatt hour, only 20 years would have been enough. Financial investments in PV, which have to fulfill the requirements of the Basel II Accord defining financial "stability" of investments, depended on the reliable statement that 25 years of operation can be expected, in this specific case. In other cases the respecitve values can be different but the general correlation is similar.

From the module manufacturer's point of view, information on the reliability of the product is also of importance for multiple reasons. On the one hand, selection/ sourcing and processing of material and equipment can be optimized using reliability information. On the other hand, warranties by nature bear a high financial risk, especially if they are as long as in PV industries. Reliability measures help to reduce the risk. In addition, if companies try to reduce their financial risk using warranty insurances, the insurance company will carefully look into the reliability of the products and also into reliability measures. In some cases, it is known that even the insurability and tariff depend on product reliability.

The importance of the above-described influence of reliability on market decisions and financial investments cannot be overestimated. Hence, proven high-quality and highly reliable modules are requested by financial investors. Supporting these developments and making reliability issues easier to understand and determine are the major goals of this book, and the authors hope to assist all interested stakeholders.

Karl-Anders Weiß, Gernot Oreški, Djamel Eddine Mansour,
Elisabeth Klimm

3 Characterization of modules and degradation effects

In this chapter, methods, technologies, and tools for the characterization of photovoltaic (PV) modules and materials are described from a quality assurance point of view as well as from a degradation analysis point of view. While this chapter focuses on the specific analytical techniques which are available for PV module and material analysis, latter chapters will give an in-depth description of loads that impact on PV modules during their lifetime (Chapter 4) as well as of the testing procedures and required equipment for accelerated aging and reliability testing of PV materials and modules (Chapters 5–7). Some tools are of use for only very specific questions or failures; in this case, the application is mentioned in the description. Other methods and tools are more of general use to analyze materials and components. A detailed description of various degradation effects and failures that can occur in PV modules is beyond the scope of this book and can be found in literature [2–6].

3.1 Destructive analytics

In this section, destructive analytical techniques are presented and explained as well as their application to analyze PV modules and materials. Destructive analytics are usually used to investigate or qualify materials before or in the production process, for example, to check the quality of incoming backsheets or the degree of cross-linking (DoC) of the encapsulation after lamination. Another application of destructive techniques is the case of specific failures which have to be analyzed or explained. Since destruction of the sample is intrinsically linked to the use of these techniques, they are not applicable to use them for answering questions on the behavior of modules in operation.

3.1.1 Gel content analysis

Gernot Oreški

Gel content analysis via Soxhlet extraction is the standard method for determining the DoC of solar cell encapsulants like EVA (ethylene-vinyl acetate) but also polyolefin elastomers. A solvent is used to leach any extractable components [7–9]. The weight of

https://doi.org/10.1515/9783110685558-003

the insoluble residue is measured after drying. The gel content or the DoC is then calculated by

$$\text{Gel content } (\%) = \left(\frac{M_2}{M_1}\right) \times 100, \quad M_2 \leq M_1 \tag{3.1}$$

with M_1 being the initial mass of the sample and M_2 the insoluble residue.

A sufficient extraction time is the key parameter for accurate and reproducible results. Also, the type of solvent and the drying conditions can influence the results [8]. Only recently with IEC 62788-1-6, an unambiguous international standard was published [7], specifying the procedure. Before that many different procedures were used in industry, which resulted in difficulties to compare gel content data from different labs and manufacturers [8]. Usually, the gel content values higher than 70% are aspired in PV module production to ensure high and constant product quality [9].

The advantage of Soxhlet extraction is the comparatively simple and well-established approach based on a gravimetric principle. However, the extraction process involves hazardous chemicals and usually takes several hours. And the sensitivity of Soxhlet extraction for poorly cross-linked samples (<30%) is very low [10]. Moreover, the procedure requires sampling and, therefore, is not capable for inline quality control. Nevertheless, it is the most common method for determining the DoC and widespread in industry.

3.1.2 Differential scanning calorimetry (DSC)

Gernot Oreški

Differential scanning calorimetry is a thermoanalytical technique that measures thermal properties of materials such as melting temperature, melting enthalpy, or glass transition temperature. In PV industry and R&D, DSC is mostly used for determining the DoC of solar cell encapsulants. In fact, DSC is also listed as a secondary method in the IEC 62788-1-6 standard [7]. The following are two approaches given for measuring the DoC: (i) the residual enthalpy method and (ii) the melt/freeze method.

Due to its simplicity, the residual enthalpy method is more common. Here the exothermic reaction enthalpy of the peroxide decomposition is measured before and after lamination process (see Fig. 3.1). The DoC X is then calculated

$$X = \frac{\Delta H_{\text{tot}} - \Delta H_{\text{part}}}{\Delta H_{\text{tot}}} \tag{3.2}$$

with ΔH_{tot} as exothermic reaction enthalpy of the uncured EVA before lamination and ΔH_{part} as exothermic reaction enthalpy of the partially cured EVA after lamination [9].

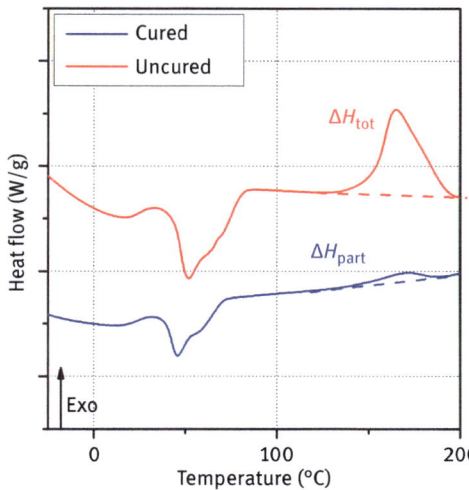

Fig. 3.1: DSC curves of cured and uncured EVA.

Next to the investigation of the encapsulant cross-linking process, DSC measurements can also be very helpful for material identification or for investigating the aging behavior of polymers [11, 12].

Chemical and physical aging mechanisms of polymers can lead to changes in melting temperature, melting enthalpy, or shifts of the glass transition temperature. An increase in the melting enthalpy of polyethylene terephthalate (PET) may indicate the so-called chemocrystallization, where short polymer chains in the amorphous phase, originating from hydrolytic chain scission, can attach to the crystalline region. This process is usually accompanied with embrittlement [12].

Also, the crystallization behavior of PET can be used as an indicator for the stage of hydrolysis. Figure 3.2 shows the DSC cooling curves for an unaged and damp heat aged TPT backsheet. After damp heat testing, an increase in the crystallization temperature was observable. Due to hydrolytic chain scission, terminal end groups are produced, which act as nucleating agents for crystallization at higher temperatures. Again, a correlation between thermal properties and embrittlement of PET has been found [11, 12]. Similar observations have also been made for backsheets using polyamide or polypropylene [13].

DSC is also useful for identification of multilayer components such as backsheets. Usually only the outer layer is accessible for identification using spectroscopic methods. The different materials can be distinguished according to their melting temperature (see Tab. 3.1).

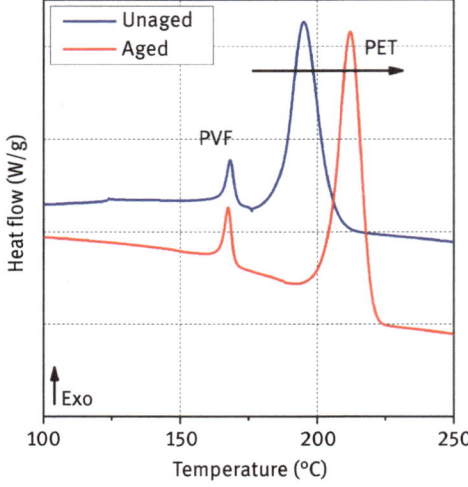

Fig. 3.2: DSC cooling curve of unaged and damp heat aged TPT backsheet.

Tab. 3.1: Melting temperatures of common polymers used in PV modules [8].

Polymer	Temperature range
EVA (encapsulant)	45–70 °C
Polyethylene/EVA (tie layers)	90–120 °C
PET	250–260 °C
PVF	190–195 °C
PVDF	160–170 °C
PA12	175–180 °C
PP	160–170 °C

3.1.3 Dynamic mechanical analysis (DMA)

Gernot Oreški

Dynamic mechanical analysis (DMA) has been used to investigate the thermome-chanical properties of encapsulant and backsheet films in the past [9, 10, 14]. Thermo-mechanical properties are strongly influenced by processing conditions, especially the cooling rate after film extrusion or PV module lamination [15]. Exposure to elevated temperatures lead to material stiffening and shift of thermodynamic transitions to higher temperatures [15], which can be measured using DMA.

Figure 3.3 shows the storage modulus E' of a common EVA encapsulant and a TPT backsheet as a function of temperature. At temperatures below 0 °C, the EVA film shows high stiffness with E' values of about 1,000 MPa. At room temperature and above the storage modulus, values decrease to about 1 MPa, giving the film its known soft character. By comparison, backsheet films are much stiffer over the whole temperature range. The TPT backsheet exhibits storage modulus values above 1,000 MPa up to 50 °C. The glass transition of PET (around 85 °C) leads to a softening, but in any case, E' values higher than 100 MPa are measured until the melting of PET at 250 °C. The PVF (polyvinyl fluoride) melting area induced only a small decrease in comparison to the glass transition of PET.

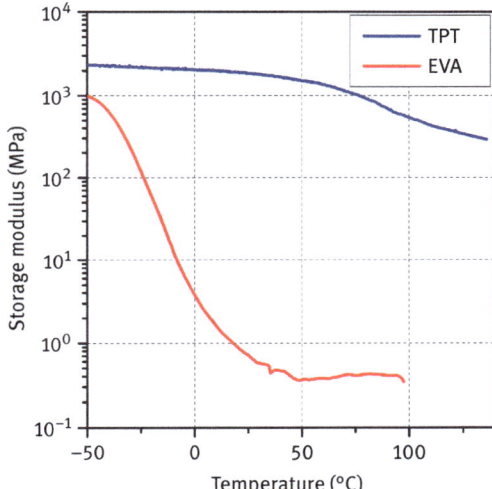

Fig. 3.3: Storage modulus as a function of temperature for TPT and EVA.

Measurements in shear mode allow for an investigation of comparatively soft encapsulant films also in the molten state. That way also the cross-linking behavior can be observed [9, 10] (see Fig. 3.4). A clear influence of the lamination time on the shear modulus of the differently cured EVA films was revealed. From 60 °C up to approximately 75 °C, the shear modulus drops, where the entire material is in the molten state. At the minimum of the modulus curve, around 125 °C, the cross-linking reaction is thermally initiated, accompanied by a significant increase in the modulus which can be attributed to the formation of a three-dimensional network. After 170 °C, the modulus levels off, becoming almost independent of temperature, which indicates the end of the cross-linking reaction. In general, compared to Soxhlet extraction and DSC, the DMA showed the highest sensitivity for determining the DoC. However, sampling and execution of the measurement are much more complicated.

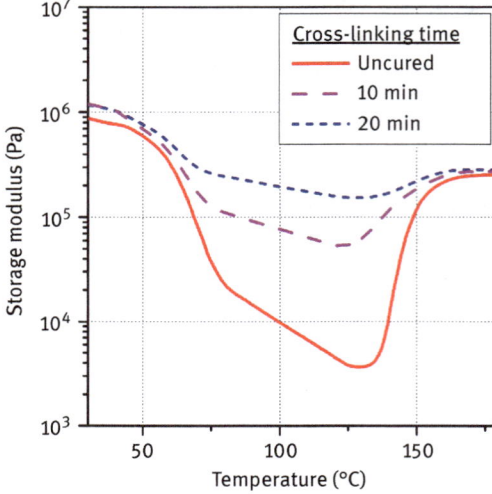

Fig. 3.4: Temperature-dependent shear modulus for an uncured and partially cured EVA film.

3.1.4 Energy-dispersive x-ray spectroscopy (EDX)

An elemental analysis of samples can be carried out by EDX (energy-dispersive x-ray spectroscopy) microanalysis, which relies on the characteristic x-ray spectrum generated by a bombardment of the sample with a focused x-ray beam. This beam of electrons induces the liberation of electrons from the inner shells of an atom which are then refilled by electrons from outer shells. This process is schematically shown in Fig. 3.5.

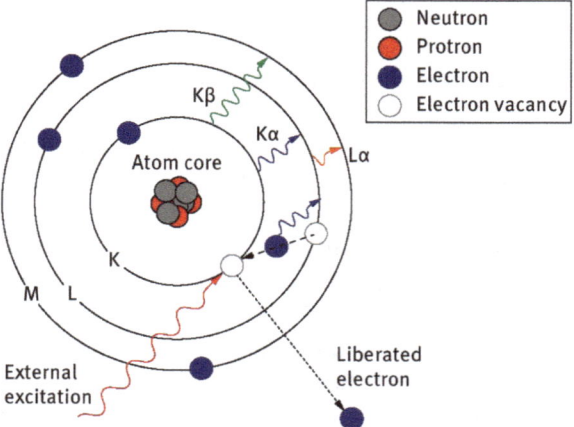

Fig. 3.5: Schematic description of electronic processes during an EDX experiment. Adapted from [16].

The Rutherford–Bohr atom model describes electrons orbiting the positive charged nucleus. In the ground state, the number of orbital electrons equals the number of protons in the nucleus, given by the atomic number Z. These orbital states have specific energies and are defined by quantum numbers. With increasing Z, orbits to be occupied aim for a minimum energy state, which have those nearest to the nucleus. Thus, they are the ones most tightly bound, which are being filled first. Orbital energy is determined mainly by the principal quantum number (n). As shown in Fig. 3.6, the shell closest to the nucleus ($n = 1$) is known as the K shell, followed by the L shell ($n = 2$) and the M shell ($n = 3$), and so on [17].

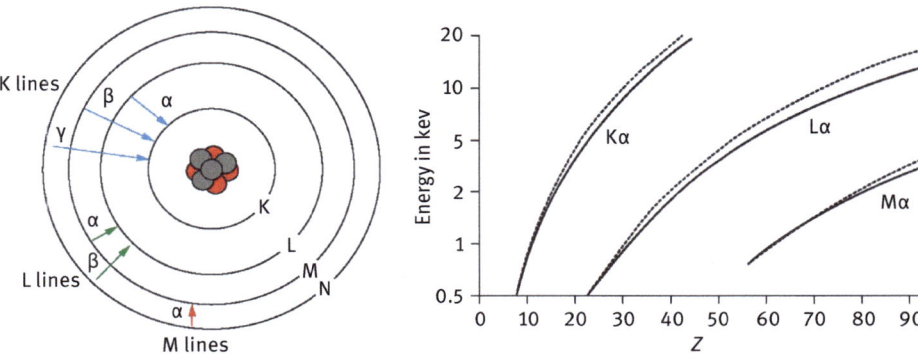

Fig. 3.6: The K, L, and M lines are generated via the electronic transition between one α, two β, or three γ shells (left). Energies of principal characteristic lines (straight lines) and their excitation energies (dotted lines) (right) are shown [18].

In PV application, the technique can be used to analyze coatings for degradation effects that occur on the surface of materials. Otherwise, sections of samples can be prepared to be analyzed by EDX.

3.1.5 Auger electron spectroscopy (AES)

Auger electron spectroscopy (AES) is a surface-specific technique utilizing the emission of low-energy electrons in the *Auger process* and is one of the most commonly employed surface analytical techniques for determining the composition of the surface layers of a sample. As shown in Fig. 3.7, the electrons that are detected by Auger spectroscopy stem from the very top of a sample, typically from the first monolayers. Thus, this method for elemental analysis of surfaces exhibits the highest surface sensitivity of the presented methods and can give a quantitative compositional analysis of the surface region of specimens, by comparison with standard samples of known composition. While the Auger electrons are generated from the

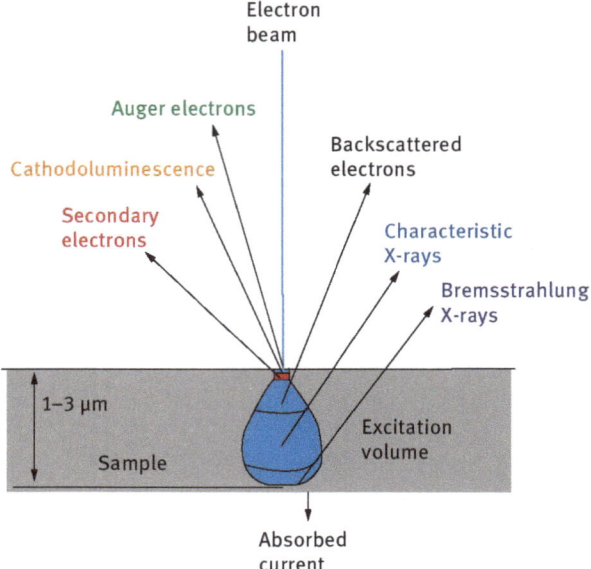

Fig. 3.7: While the Auger electrons are generated from the first monolayers of a sample, cathodoluminescence, secondary electrons, and x-rays stem from deeper layers.

first monolayers of a sample, cathodoluminescence, secondary electrons, and x-rays stem from deeper layers.

In PV, the tool can be used to analyze coatings of degradation effects that occur on surfaces of materials. Otherwise, sections of samples can be prepared to be analyzed by AES. It is also possible to analyze deeper layers (still in the µm range) with AES by applying reverse sputtering.

3.1.6 Peel testing

Peel testing is used in PV to determine the adhesion of interconnector ribbons to solar cell metallization, but also for the adhesion of the other PV module components to each other. Peel testing can be done using standard material testing machines but there are also adapted setups available specifically for the use in PV as shown in Fig. 3.8. These adapted setups enable tests of the adhesion of backsheet materials to the encapsulation (Fig. 3.9), adhesion of the encapsulant on the cell, or the quality of the soldered or glued connection of cell interconnectors to the cells.

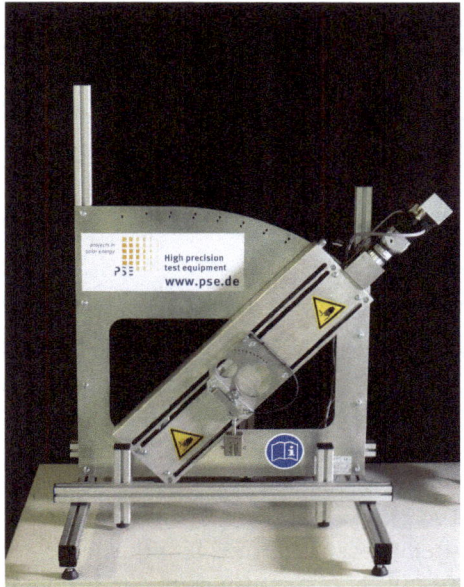

Fig. 3.8: Special peel testing device for PV applications with adjustable peel angle. Source: PSE AG.

Fig. 3.9: Peel testing to analyze adhesion of backsheet to encapsulation. Source: PSE AG.

3.2 Nondestructive analytics

In this section, nondestructive analytical techniques are presented, which have the advantage of functional capability without extensive and invasive sample preparation that allows the modules to stay intact, for additional testing or further operation. For most questions related to reliability, changes of PV module properties or parameters have to be analyzed. This is much easier if measurements can be done at the same sample after different (aging/exposure) steps, since the influences of differences between samples are excluded. Due to that fact and since testing especially of full modules is very time-consuming and so also costly, nondestructive methods are preferred and used whenever they are available.

3.2.1 IV curve measurements

The power output of PV modules is typically determined by a measurement under standardized testing conditions (STC) at 25 °C, with 1,000 W/m^2 irradiation and an air mass 1.5 spectrum as by the standard IEC 60904-5. To deliver comparable results, the equipment has to fulfill very high requirements regarding spectrum, time stability, and lateral homogeneity. The quality of a setup regarding these three parameters is described by quality classes. Good equipment should reach class "A" in all parameters, to be listed as "AAA." In most cases, the measurement is performed in a flasher tester using a very accurate light flash (Fig. 3.10) but there are also setups using solar simulators providing steady-state illumination. These setups are called DC or steady-state simulators.

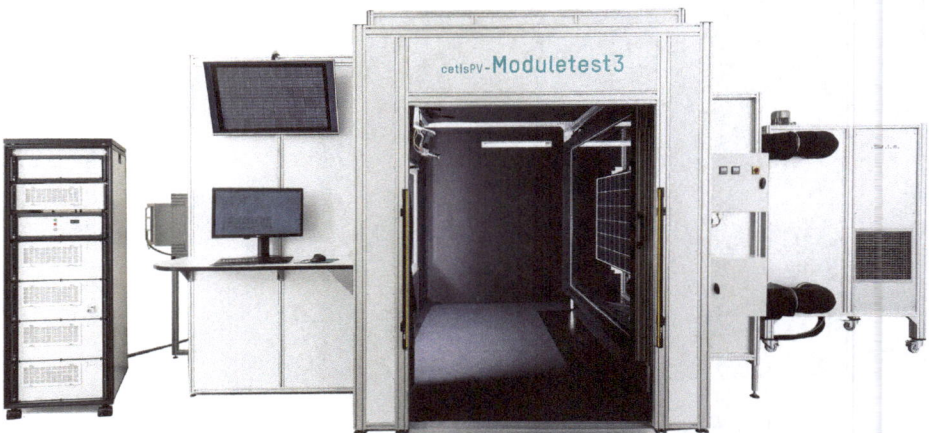

Fig. 3.10: State-of-the-art flasher setup for module analysis with control and operation unit (left) and temperature control (right). Source: h.a.l.m. GmbH.

Since power is the economically most relevant parameter of a PV module, the IV curve measurement is of high relevance. The maximum power is calculated at the point of the IV curve where the value of $I \times U$ is maximal (maximum power point (MPP)) and is therefore called P_{MPP} (Fig. 3.11).

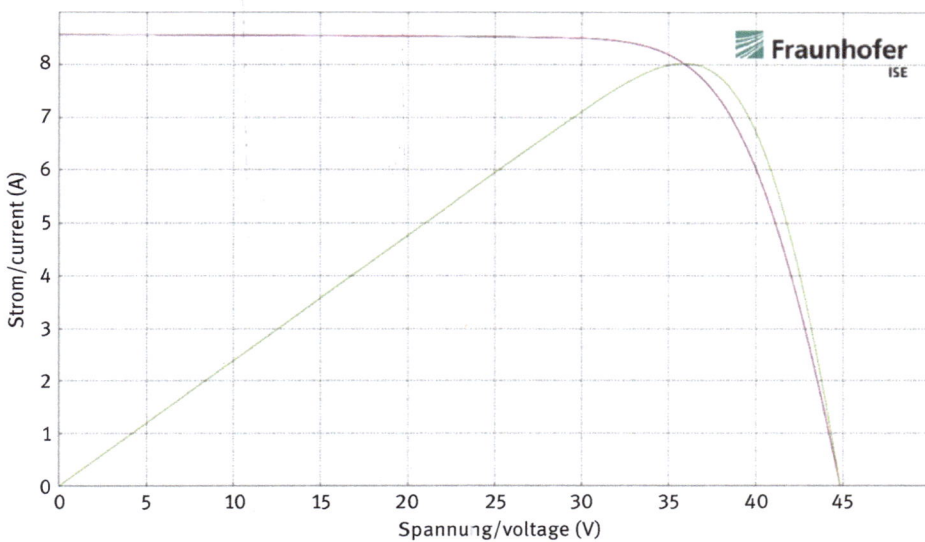

Fig. 3.11: Standard IV curve of a PV module showing the relation of I and U values (red). The green line represents the calculated power $P = I \times U$ and so also P_{MPP} at its maximum.

Degradation is usually defined by the loss of performance as well as the end of the (economic) service life, which is often defined by a certain loss of performance, typically when the performance has reached a certain level of e.g. 80% of the initial or name plate performance. Also pass–fail criteria of tests or the fulfillment of conditions of contracts are often defined by performance data. These contracts often allow a defined level of minimum power or a variation of the labeled power. Especially the given warranties in PV usually only define power values.

The power performance of a module, on the one hand, integrates all kinds of effects, which have taken place within a module. Degradation effects of materials are therefore also to be seen as an influencing parameter. On the other hand, performance data alone cannot explain specific effects or processes since it cannot be directly linked to specific components or processes. The complete IV curve is much more powerful and allows determining more electrical parameters of the module, like short-circuit current I_{sc}, open-circuit voltage U_{oc}, series resistance R_s, parallel resistance R_p, and fill factor, which can be easily linked to the effects in specific materials or components. The form of the IV curve also allows the identification of some degradation or operational effects (Figs. 3.12 and 3.13).

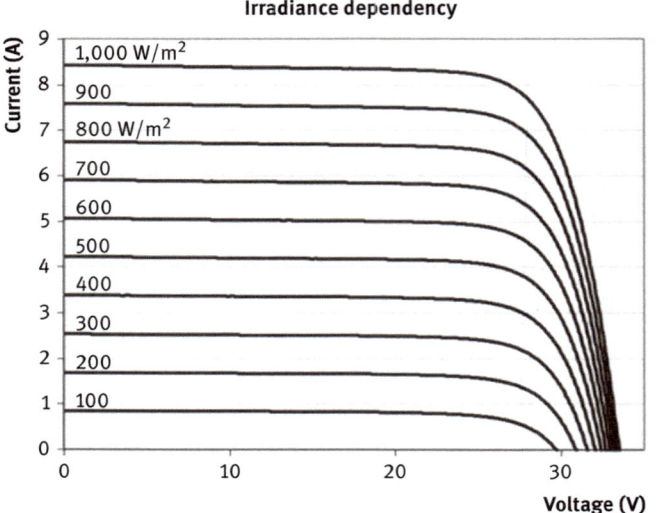

Fig. 3.12: IV curves taken of one sample under different irradiation levels. It can be seen that irradiation mainly has an impact on the current. This allows to identify e.g. effects of reduced transmission of glazing or encapsulation of a module.

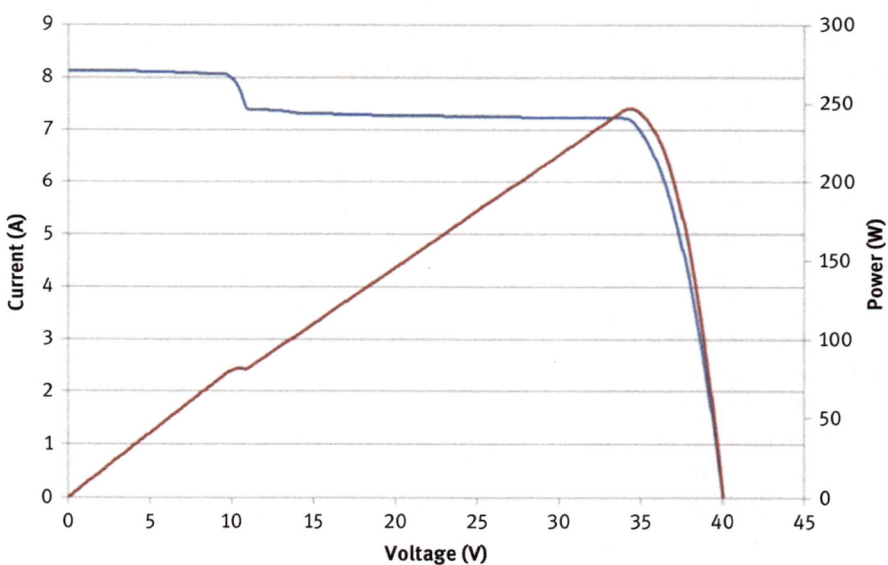

Fig. 3.13: IV curve showing a step, indicating an issue, for example, partial shading.

3.2.2 Electroluminescence (EL) imaging

Fully processed solar cells or cells in modules can be excited by application of a voltage leading to recombination effects in the semiconductor and so to the emittance of radiation in the IR range, allowing for electroluminescence (EL) imaging. Due to its simplicity, modules only have to be connected using the standard cabling, and an image has to be taken using an IR camera (compare Section 3.2.12) in dark environment. EL imaging is nowadays widely used for module inspection (Fig. 3.14). The dark environment is useful to decrease the background "noise" during the EL imaging but not absolutely necessary. EL images can even be taken outdoors using lock-in technology or adapted filters.

It is used to analyze cells and cell strings before lamination, as well as modules at the end of production lines for quality assurance purpose. EL is used as test for incoming modules after shipment and also for assessment of degradation effects. EL setups are often directly combined with the IV curve measurement installations to perform combined tests.

Fig. 3.14: Setup for EL imaging. Source: Greateyes GmbH.

EL intensity is, among other factors, limited by the conduction properties of the solder bonds and the grid. In other words, a corroded or insufficient adhering metallization onto the emitter reduces the EL signal drastically and so can also be detected by EL. Another major use of EL is to detect cell cracks and breakage. Cell cracks appear as dark lines on the solar cell in the EL image and broken parts simply remain dark. It has been shown that some cracks, especially if they are of limited depth, can only be seen under specific voltage levels. This means that it is recommended to take EL images using different voltages and not only the maximum voltage to gain a complete impression of the status of the cells and the electrical connection. The interpretation of EL images can be automized to a limited degree. Deep cracks, especially cell break-age, or bad grid fingers are easy to be detected by image analysis tools (Fig. 3.15), but detection of further effects by experienced operators is usually more successful.

Fig. 3.15: EL image of full-size PV module taken at 39.4 V showing cracks in cells and also broken sections.

3.2.3 Internal and external quantum efficiency (IQE) and (EQE) measurements

The external quantum efficiency (EQE) can be defined as the ratio of the number of charge carriers collected by the solar cell to the number of photons of a given energy reaching a solar cell (incident photons). The internal quantum efficiency (IQE) is de-fined as the ratio of the number of charge carriers collected by the solar cell to the number of photons of a given energy that reaches a solar cell from outside and is absorbed by the cell.

The measurement of the EQE and IQE allows for an evaluation of the spectral re-sponse of a solar cell. The setup of an EQE measurement system used at the Fraunhofer ISE is shown in Figs. 3.14–3.16.

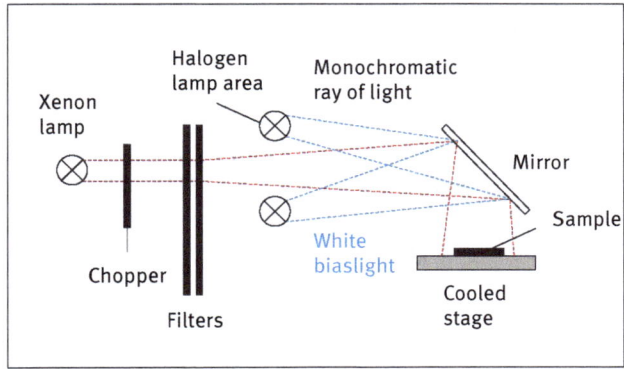

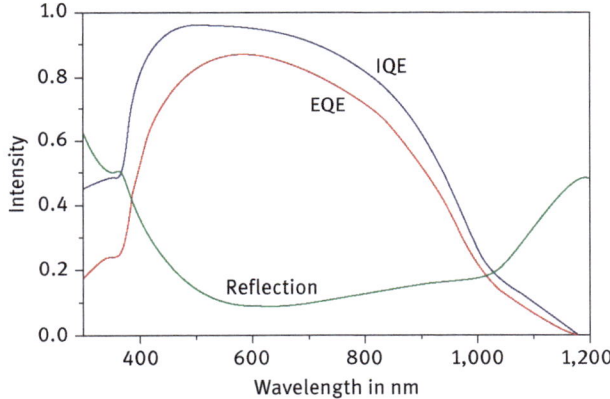

Fig. 3.16: Upper figure: a schematic description of the setup of an EQE measurement system utilizing a xenon arc lamp as light source is shown. Typical curves of the IQE and EQE as well as the reflection of a solar cell are given in the lower graph [19].

A xenon arc lamp can be used as the radiation source. The wavelength selection is carried out using an optical filter, and the light modulation is performed by a chopper. Typical curves of the IQE and EQE and the reflection of a solar cell as shown in Fig. 3.16 demonstrate that the IQE and the EQE are connected via

$$IQE - EQE \frac{1}{1-R} \tag{3.3}$$

where R is the reflection of the solar cell.

3.2.4 Photoluminescence (PL) imaging

Luminescence imaging of silicon samples utilizes the luminescence emitted from the surface of a sample as a result of radiative recombination of electrons. These electrons stem from a previous excitation. In addition to EL imaging described earlier, photoluminescence (PL) imaging can be used for PV module characterization with a focus on the solar cell properties. The difference between these two techniques is the way of luminescence excitation: While PL imaging uses a UV light source, EL utilizes voltage. Due to these different excitation processes, the factors influencing the efficiency of electron generation are different. Figure 3.17 displays the setup of a PL stage consisting of a widened laser beam for the solar cell excitation. The resulting luminescence is detected by a camera with a lock filter that decouples the reflected laser light. PL imaging uses optical excitation, which allows application to a wider range of samples, including bricks, as-cut wafers, and partially processed wafers. The PL intensity is mainly reduced by impaired silicon cell properties such as a deteriorated AR coating or emitter region.

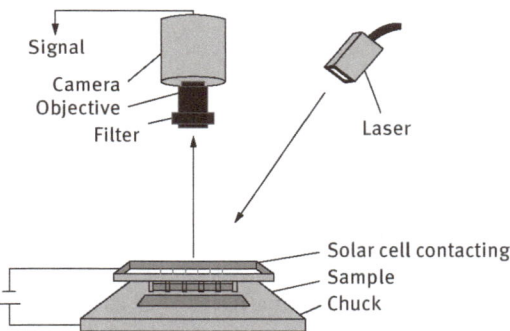

Fig. 3.17: Setup of a PL stage consisting of a widened laser which excites the solar cell. The resulting luminescence is detected by a camera with a lock filter that decouples the laser light which is reflected by the sample [19].

3.2.5 Gloss and color measurement

Yellowness can be defined as the deviation in chroma from whiteness in the dominant wavelength range from 570 to 580 nm. The Yellowness Index (YI) is the magnitude of yellowness relative to a white standard under illumination by a standard light source [20].

The YI is a measure of quantifying the subjective perception of colors by the human eye. This index is most commonly used to evaluate color changes in a material, for example, caused by outdoor exposure or accelerated aging.

It can be determined via different standards given by the Commission Interna-tionale de l'Ecleirage (CIE) or by the American Society for Testing and Materials (ASTM). The determination of the YI according to the CIE standard will be discussed in the following and is visualized in Fig. 3.18.

Three things are crucial for the perception of a color:
- the light source,
- the investigated object, and
- an observer.

Since this perception of a color highly depends on the character of the selected light source, illuminants are standardized according to their color temperature.

Some of the most important standard illuminants are
- A (incandescent, 2,856 K),
- C (average daylight, 6,774 K),
- D65 (daylight, 6,504 K),
- F (fluorescent).

The color temperature of a specific illuminant is determined by the spectral distribu-tion of its radiation energy. The chromaticity coordinates can be directly obtained via a chromaticity diagram, which is shown in Fig. 3.18. Due to the relationship

$$C_X + C_Y + C_Z = 1 \tag{3.4}$$

only two chromaticity values are needed for an unambiguous assignment; thus, chromaticity is usually plotted in a 2D diagram.

Mostly C or D illuminants with a color temperature above 6,000 K are utilized for spectroscopic measurements. In addition to the character of the illuminant, the spectral curve of the investigated object and the sensitivity curves of the human color receptors have to be considered.

A specific color consists of proportions of the variables X, Y, and Z, which are called the tristimulus values. The human eye perceives color via three receptors, which detect the proportions of the light. Those receptors primarily detect the red-dish purple ($\bar{x}$), the green ($\bar{y}$), and the blue ($\bar{z}$) components of the light. The sensitiv-ity distribution of the receptors depends on the angle of incidence of the light. For calculations either a 2° (2° observer) or a 10° (10° observer) angle is considered. To quantify the human perception of a light stimulus by an object, the product of these values and the energy distribution E_λ of the sample stimulus, that is, reflected, transmitted, or emitted light, are integrated over a wavelength range as shown in eqs. (3.5)–(3.7). This way, the tristimulus values X, Y, and Z are obtained:

$$X = \int_{380}^{760} \varphi_\lambda \bar{x} \tag{3.5}$$

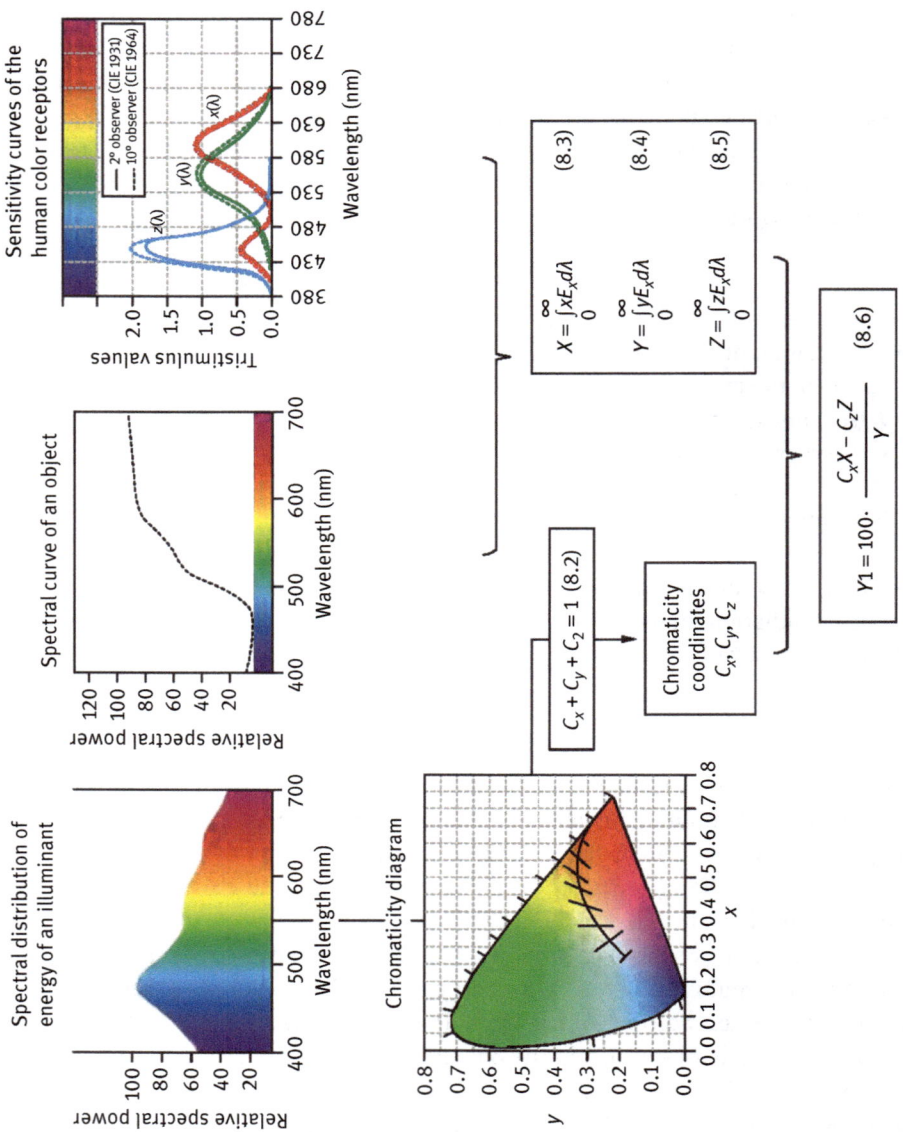

Fig. 3.18: The determination of the Yellowness Index is accomplished via the calculation of the chromaticity coordinates of the illuminant and the calculation of the tristimulus values, taking the spectral curve of the investigated object and the sensitivity curves of the human color receptors into account [21].

$$Y = \int_{380}^{760} \varphi_\lambda \bar{y} \qquad (3.6)$$

$$Z = \int_{380}^{760} \varphi_\lambda \bar{z} \qquad (3.7)$$

The chromaticity coordinates can be directly obtained via a chromaticity diagram, which is shown in Fig. 3.18 or calculated according to DIN 5033 with eqs. (3.4)–(3.8):

$$YI = 100 \left(\frac{C_X X - C_Z Z}{Y} \right) \qquad (3.8)$$

where $C_X = 1.28$ and $C_Z = 1.06$ [84].

Smaller values of YI indicate lower yellowish color. Color differences can be perceived by the human eye if ΔYI is in the range of 5 or bigger.

Gloss is an optical impression of a surface that is physically defined as the ratio of the direct and the diffuse reflected light incident on the surface. There are gloss meters available, which are easy to handle and directly deliver the gloss value. Descriptions and definitions of the relevant parameters and procedures can be found in the standard ISO 2813. The loss of gloss is often an early and easy detectable indicator for degradation effects, especially for surfaces and coatings.

3.2.6 Raman spectroscopy

The Raman effect originates from the interaction between light and matter, resulting in a wavelength shift of the scattered compared to the incident light.

Light interacting with matter may either be absorbed, scattered, or transmitted. Absorption only occurs when the energy gap between the molecule's ground and excited states matches the energy of the incident photon. In this case, the molecule is promoted to an excited state, and the loss of energy can be detected to quantify the amount of absorption. With regard to the scattered light, the wavelength of the major part of the scattered light is identical to that of the incident light. But a minor part exhibits a shift of the wavelength which is induced by molecular vibrations and rotations – this phenomenon is referred to as Raman scattering.

A first understanding of the Raman effect can be obtained from energy scheme considerations, which are visualized in Fig. 3.19.

If a molecule is hit by a photon of the energy $E = h v_0$ of an energy smaller than $E(S_0 \rightarrow S_1)$, the molecule is excited from the ground state S_0 to a virtual level. This molecule can thereafter either relax into the ground state, emitting a photon with the initial energy $h v_0$ (Fig. 3.19: blue lines) or into an excited vibrational level under

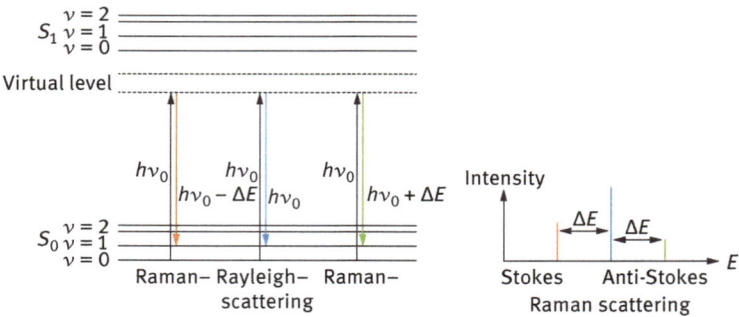

Fig. 3.19: Energy scheme and frequency spectrum showing the Raman and the Rayleigh scattering. Left: Electronic transitions between vibration levels of the ground state S_0 and the virtual level, resulting in Raman and Rayleigh scattering. Right: The intensities of the Stokes and the anti-Stokes scattering, which exhibit an energy shift of the same amount with respect to the initial energy, are small compared to the Rayleigh scattering. The intensity of the Stokes scattering is higher than that of the anti-Stokes scattering due to the occurrence of fewer molecules in an excited vibrational state [21].

emission of a photon with a reduced energy $h\nu_0 - \Delta E$ (Fig. 3.19: orange lines). Here, the first case describes an elastic scattering process – the Rayleigh scattering. The second case describes the inelastic Stokes–Raman scattering. In addition to the Stokes–Raman scattering, a second inelastic scattering process can occur under the prerequisite of molecules in an excited vibrational state: The anti-Stokes–Raman scattering. Anti-Stokes–Raman scattering occurs when initially excited molecules are further excited into a virtual level and relax into the ground state emitting a photon of the energy $h\nu_0 + \Delta E$. As shown in Fig. 3.19 (right), the intensity of the Stokes scattering is under ambient conditions much higher than that of the anti-Stokes scattering due to the small amount of molecules in an excited vibrational state. At room temperature, most molecules are in the lowest vibrational energy level. But at higher temperatures, the probability of molecules being in an excited state, which is given by the Boltzmann distribution, rises and scattering can also occur from the excited to the ground state.

An example of a sophisticated Raman microscope – which is in this case also combined with an atomic force microscope (AFM) in order to allow surface property analysis – is given in Fig. 3.20.

Raman scattering is, compared to Rayleigh scattering, a very weak process with only one in $10^{-6} \ldots 10^{-8}$ photons showing Raman scattering. Thus, the intensities of the Stokes and anti-Stokes scattering are small compared to the Rayleigh scattering.

An often observed phenomenon is the occurrence of a fluorescence background in the Raman spectra. This fluorescence occurs when the excitation energy is high enough to promote the molecule to an excited singlet state (S_1, S_2, . . ., S_n) as shown in the Jablonski scheme in Fig. 3.21. This excited molecule immediately falls down to the S_1 state via internal conversion (IC). From this state, the luminescence processes

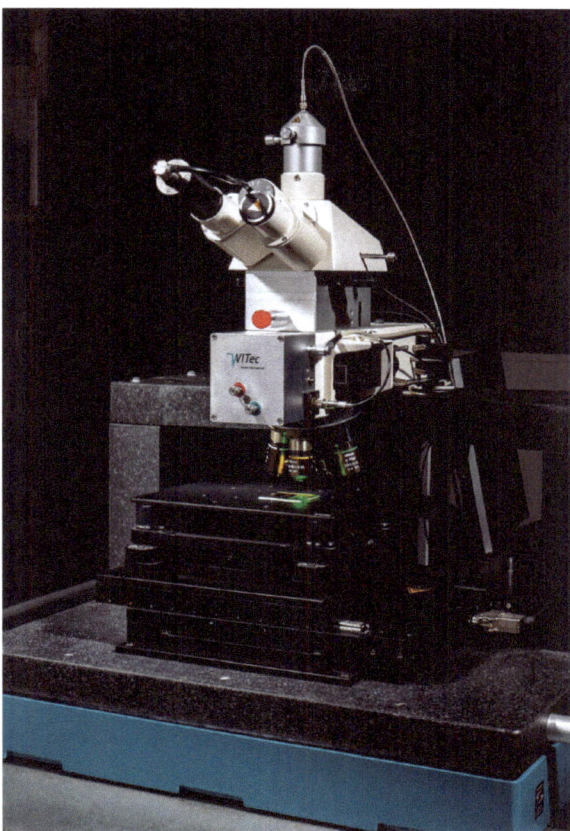

Fig. 3.20: Combined Raman microscope and atomic force microscope.

fluorescence or phosphorescence may occur. In case a triplet state, obtained via intersystem crossing (ISC), is included, subsequent phosphorescence occurs. In contrast to fluorescence, phosphorescence involves a change in the spin multiplicity. Therefore, this process is relatively slow due to the forbidden $S \to T$ transition as stated by the Frank Condon principle. Thus, phosphorescence processes exhibit longer lifetimes compared with fluorescence processes. The relative importance of fluorescence compared to phosphorescence is determined by the rates of the IC and the ISC processes [22].

The potential fields of application of Raman spectroscopy in polymer characterization are diverse. Besides material identification, including copolymer composition and structure analysis, a conformational analysis can be carried out.

Information about the polymerization kinetics and mechanisms can also be obtained with little sample preparation effort, and also an in situ characterization is possible. This polymerization analysis is mostly based on the observation of the decrease in the C=C bond vibration of the reactant.

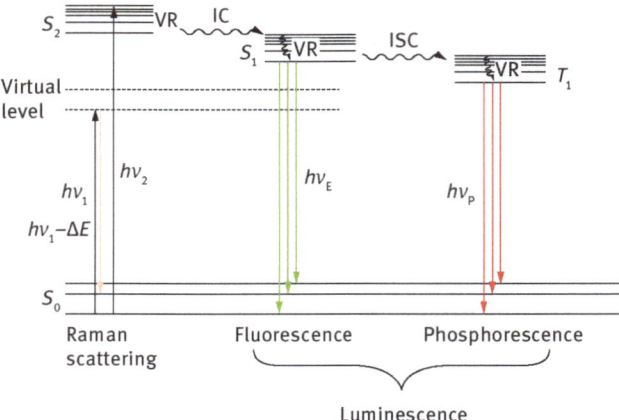

Fig. 3.21: The Jablonski scheme describes the electronic transitions of internal conversion (IC) and intersystem crossing (ISC), involving excited singlet (S) and triplet (T) states, which lead to the luminescence processes fluorescence and phosphorescence in contrast to the Raman scattering [21].

Furthermore, the degree of crystallinity of polymers can be determined via certain vibrations characteristic for the amorphous or the crystalline phase. An example is the crystallinity determination in polypropylene [23] or determination of the cross-linking of EVA [85].

Different additive types are added to the polymer formulation in order to enhance processing properties, material properties of the prospective component, or polymer stability. The analysis of those additives within the polymer is of great importance in order to identify the quality of the additives, their dispersion across the polymer, and so on. Raman spectroscopy allows the simultaneous identification of various additives in parallel. By using Raman microscopy, even the spatial dependency of the additive distribution can be obtained [24]. In PV, Raman spectroscopy can also be used to analyze the DoC of the encapsulation as this can be analyzed by determining the ratio of double and single bondings.

Additionally, Raman spectroscopy can be used to monitor polymer degradation in a nondestructive and quick manner. Since most polymer degradation reactions can be traced back to thermo- or photo-oxidative processes, this Raman spectroscopic degradation analysis is often the observation of the occurrence of peaks of oxidized species [25].

In order to analyze full-size PV modules, a Raman probe can be attached to the Raman spectroscope. In the example given in Fig. 3.22, the Raman probe is attached to a linear motor, which allows an automatic movement of the probe and so also reproducible measurements.

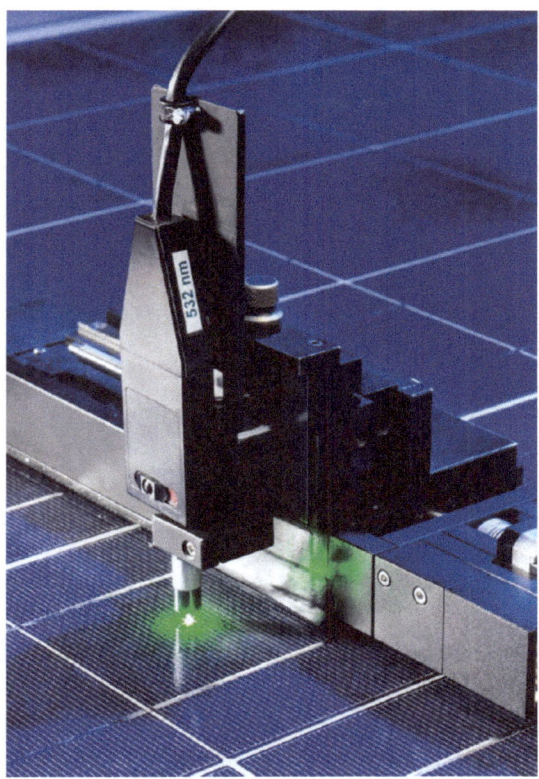

Fig. 3.22: A Raman probe which is attached to a linear motor.

3.2.7 Nanoindentation

Djamel Eddine Mansour

Nanoindentation has been recently used as a suitable method to investigate the changes in viscoelastic properties of the polymeric components in PV modules (backsheet and encapsulant) occurring during the lamination process and accelerated aging. Using mechanical characteristics, including hardness, modulus, load/displacement profile, creep hold response, and residual impression, the instrumented nanoindentation was able to distinguish between different PV backsheets and quantify the effects of accelerated testing [25]. For the encapsulant, additional time-dependent indentation cycles are applied using a spherical or flat punch tip (i.e., dynamic mode using frequency sweep [26] or creep testing mode [27]). Those modes were used as a fast and simple method for DoC determination by measuring quantitative mechanical parameters of the encapsulant. Figure 3.23 illustrates the nanoindentation damping factor (tan δ) for the EVA surface as a function of DoC for different tested

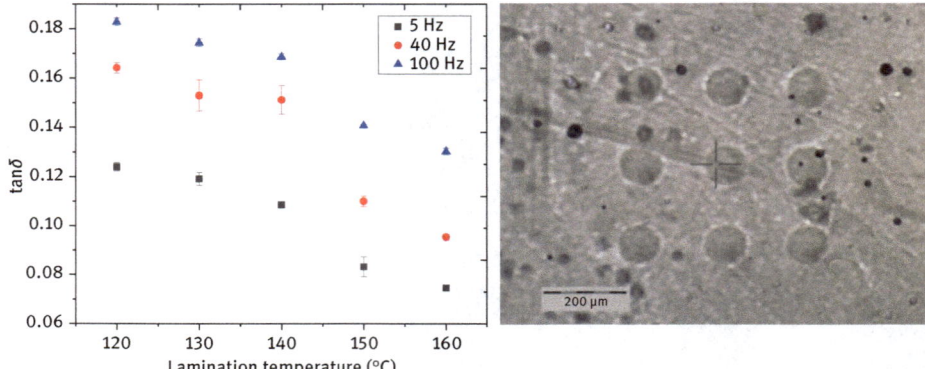

Fig. 3.23: Left: Damping factor (calculated from the sinus part of the dynamic curve) of the EVA surface as a function of DoC as measured by DSC. Right: Optical microscopy image of indented EVA surface (glass–EVA interface). Source: Djamel Eddine Mansour.

frequencies. Each point is an average of nine indentation measurements at a given frequency. Five samples with different lamination temperatures were used for this experiment. DoC samples have been determined by DSC.

A decrease in tan δ corresponds to an increase of cross-linking bond density, which also increases the molecular weight of the material. The strongest decrease of tan δ was found between the values corresponding to lamination temperatures of 140 °C (48% DoC) and 150 °C (67% DoC). At higher temperatures, a slight leveling off of the values is shown, which would indicate thermomechanical stability. Nevertheless, the same trend for tan δ is observed at all frequencies; the higher the DoC the lower the tan δ. Furthermore, similar values of tan δ of the same EVA samples and the same frequencies could be observed by using DMA as a conventional method.

The effect of the backsheet's properties on the encapsulant surface degradation can also be investigated using nanoindentation. To this end, different backsheets were used to understand their influence on encapsulant degradation rate during UV–DH combined aging. Additional characterizations were performed to interpret the indentation measurements, including a chemical structure study on the same surface using Fourier-transformation infrared (FT-IR) spectroscopy, and phase-transition measurements using DSC [28].

3.2.8 Fourier-transformation (FT-IR/UV/vis) spectroscopy

The transmission and reflection properties of PV components or PV laminates can be quantified by means of FT-IR/UV/vis spectroscopy analyzing different parts of the spectrum, ultraviolet (UV), visible (vis), or infrared (IR).

The general correlation of transmission (T), reflection (R), and absorption (A) is given as follows:

$$T + R + A = 1 \tag{3.9}$$

The absorption of light is described by Lambert–Beer's law:

$$A_\lambda = \lg(I_0/I) = \varepsilon_\lambda c d \tag{3.10}$$

where A_λ is the absorbance, I_0 is the intensity of the incident light, I is the intensity of the transmitted light, ε_λ is the extinction coefficient, c is the concentration of the absorbing species, and d is the path length.

Since the absorbance of a sample is related to the transmittance T_λ via an inverse logarithmic dependency

$$A_\lambda = \lg(1/T_\lambda) \tag{3.11}$$

the amount of light transmitted through a certain sample can therefore be calculated according to

$$T_\lambda = I_0/I = 1 - R_\lambda - A_\lambda \tag{3.12}$$

where R_λ is the reflectance and of the sample.

For FT-IR/UV/vis spectroscopic measurements, light of a broad spectral range is employed for the excitation of molecular vibrations in the IR, UV, and visible range. With different light sources such as xenon lamps, tungsten lamps, metal halide lamps, and SiC globars, a wavelength range between 200 and 17,000 nm can be achieved. For the detection of the transmitted or reflected light, different detectors, such as photomultipliers, silicon detectors, indium-gallium-arsenide (InGaAs) detectors, and mercury cadmium telluride (MCT) detectors, can be used.

The measurement by the use of larger aperture areas ($A = 2$–5 cm^2) is recommended due to an often inhomogeneous nature of sample surfaces. Reproducible and comparable measurements of not completely uniform materials/surfaces can be accomplished using integrating spheres (Ulbricht spheres), which are shown in Fig. 3.24. Ulbricht spheres consist of a hollow spherical cavity with a diffuse reflective coating, such as white barium sulfate for the UV/vis range or gold for the IR range, and small holes for entrance and exit ports. Incoming light is homogenously distributed within the integrating spheres. Therefore, an integrating sphere is a diffuser which preserves information on T, R, or A but eliminates spatial information.

The specific detectors for the UV, the visible, and the IR range are situated on small holes of the sphere. The measurement can be performed in a transmission mode (right) or reflection mode (left), depending on the mounting of the sample behind or in front the sphere. Integrating spheres can, among other things, be utilized for the measurement of the radiation characteristics of light sources or to measure the diffuse reflectance of surfaces, including an average over all angles of illumination and observation.

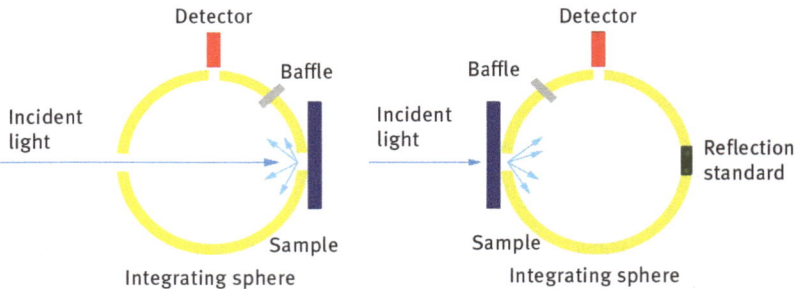

Fig. 3.24: Integrating spheres are either utilized in reflection (left) or transmission mode (right). The incoming light is scattered by the diffuse reflective coating on the inside of the sphere. The detection of the scattered light is achieved via detectors for the different spectral ranges.

3.2.9 Scanning acoustic microscopy (SAM)

Djamel Eddine Mansour

Scanning acoustic microscopy (SAM) is a nondestructive technique that uses focused sound for the investigation of a sample. This way, voids, cracks, and delaminations within microelectronic packages can be detected in PV modules. The technique is based on acoustic signals from the sample. Therefore, a microscope produces, transmits, and detects short pulses with a high penetration rate (Fig. 3.25). Changes in the partial transmission and reflection of ultrasound can be detected in case of changes in the acoustic impedance, which occur if material boundaries or property changes are present. Frequencies up to 2,000 MHz are used for the detection of potential problems within the laminate. SAM offers the possibility of analyzing the compliance of PV module components because it can provide real-time acoustic high spatial resolution information from the internal structure of materials [29], for example, the measurement of the longitudinal modulus of backsheet and EVA foil inside a PV module. As a validation of SAM, it was shown that the longitudinal modulus in the backsheet and the EVA have a similar trend with temperature. Moreover, the change in the modulus of these materials at 15 MHz has a good correlation with the Young's modulus calculated from the tensile tests [30].

Furthermore, it was reported that SAM delivers complementary information to EL imaging and dark lock-in thermography on the quality of PV modules [31]. Moreover, SAM was proved to be in good agreement with EL [32].

As an example, Fig. 3.26 shows a SAM image of the EVA–cell interfaces, where the red dashed squares indicate bright structures along the silver (Ag) pads. Additionally, the blue dashed square indicates a large bubble on top of the cross-connector. The bright spots show voids that were caused by excessive release of volatile organic compounds during lamination [33]. The acoustic cross section confirms that the voids are located at the EVA–cell interface, and appear slightly earlier than the cell signal.

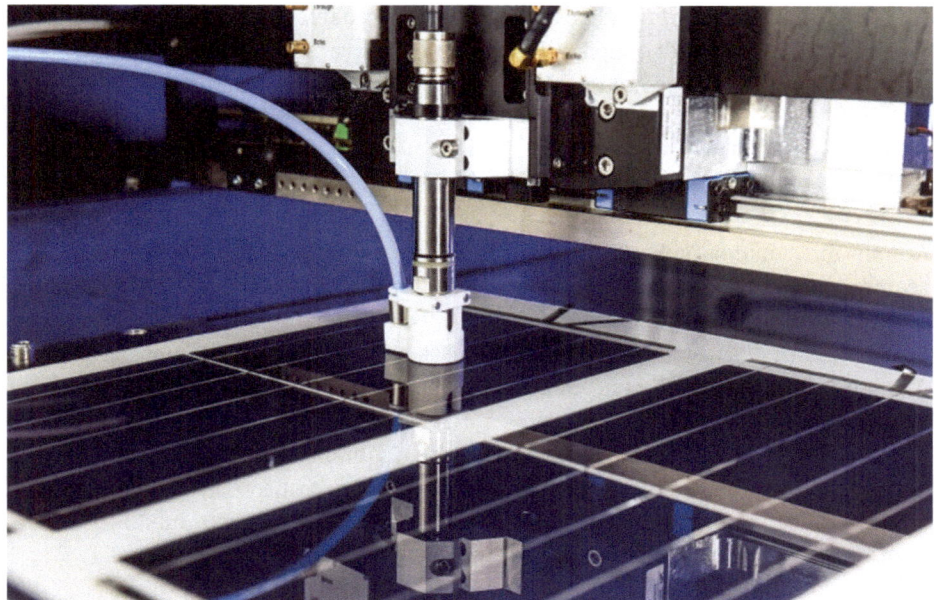

Fig. 3.25: PV module being scanned from the front side by SAM. Source: Djamel Eddine Mansour.

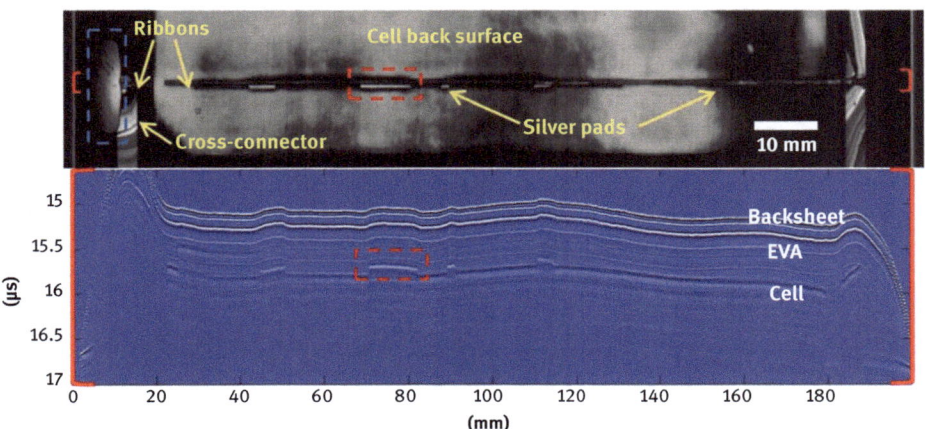

Fig. 3.26: Top: SAM image of a PV module with postlamination defects: at the backside of solar cell $f = 30$ MHz, gate length: 80 ns; resolution: 80 µm/pix. Bottom: Acoustic cross-sectional view depicting the voids along the Ag pads. The red line in the top acoustic image indicates the cross-sectional plane. Source: Djamel Eddine Mansour.

Toward this goal, degraded PV modules are analyzed by SAM to understand and localize failure modes such as backsheet cracking in specific layers and delamination that are otherwise not visible without destroying the module.

3.2.10 Atomic force microscopy (AFM)

Atomic force microscopy (AFM) is a very-high-resolution surface-sensitive type of scanning probe microscopy, with demonstrated resolution on the order of fractions of a nanometer, which is more than 1,000 times better than the optical diffraction limit. Unlike light or electron microscopes, AFM uses a tactile approach to image a surface. A small probing tip at the end of a flexible cantilever scans very closely across the sample surface, so that atomic range forces act between the tip and the sample (Fig. 3.27). Typical AFM setups are mounted on a vibration isolated substructure and are optimized for high precision sample analysis in controlled conditions (Fig. 3.28). They are usually limited to samples of small size and lower weight due to the limitations of the setup regarding size, and the scan table regarding weight.

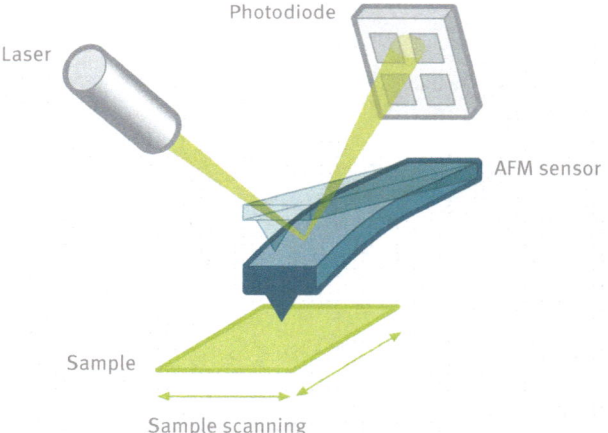

Fig. 3.27: Principle of AFM measurement. The probe (cantilever) consists of a beam with a very small tip which is moved over across the sample and the deflection is monitored. Source: WITec GmbH.

There are different measurement modes used for AFM measurements (Fig. 3.29), generating different kinds of information on the surface properties and also having different impact on the surface. While the contact modes can potentially cause mechanical effects on very sensitive surfaces, this is not really relevant in PV; the non-contact modes are not directly impacting the surface.

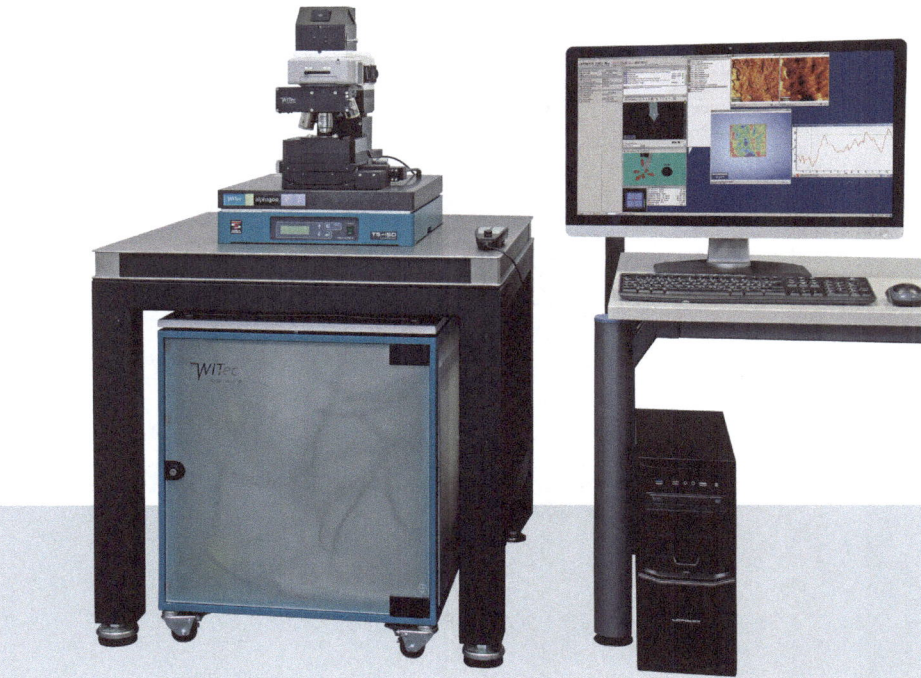

Fig. 3.28: Typical AFM setup for analysis of smaller samples including control electronics and data visualization. Source: WITec GmbH.

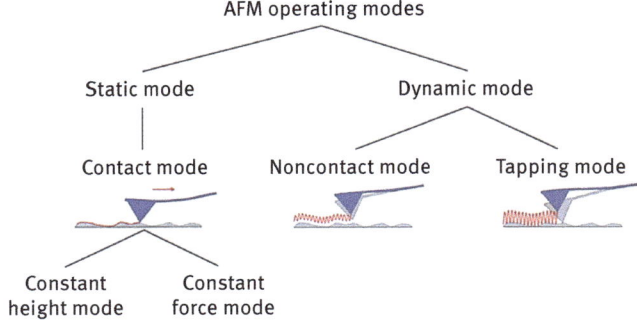

Fig. 3.29: Overview on the typical AFM measurement modes. Source: Johanna Banken.

Measurements with the dynamic noncontact mode called tapping mode are very popular since they are very sensitive and generate additional data in one scan besides the topography data. In the tapping mode, also data on the amplitude of the oscillation of the cantilever and the phase shift of the cantilever relative to the excitation is determined (Fig. 3.30). Amplitude and phase shift data can deliver additional

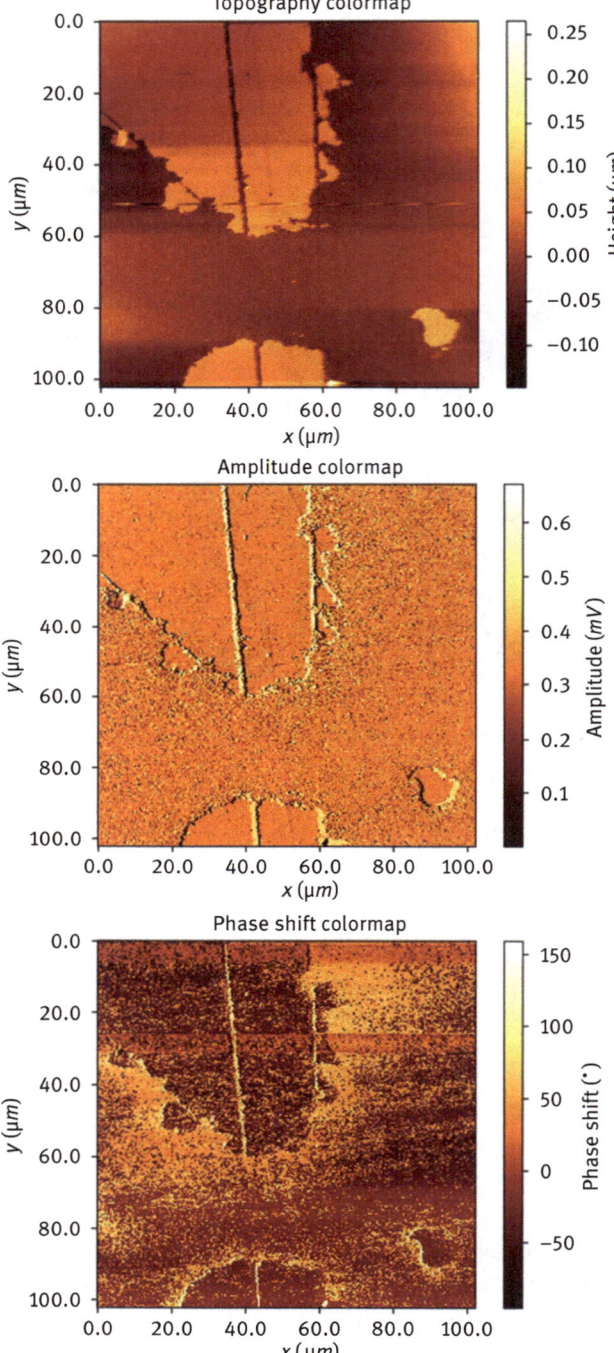

Fig. 3.30: Images showing AFM data of one sample of solar glass with coating. The data on topography, amplitude, and phase shift have been generated in one tapping mode measurement with a scan size of 100 × 100 μm². The different surface properties of the images show that the coating obviously has been degraded and already ablated in some areas. Source: Luis Armando Blanco Bohorquez.

information on the sample surface properties, for example, related to the elasticity or adhesive properties.

There are also further AFM measurement modes for more complex or adapted surface analysis. One powerful example is the so-called pulsed force mode (Fig. 3.31), which combines basic features of dynamic AFM measurement with nanoindentation (Section 3.2.7) using lower frequencies than dynamic AFM modes and lower forces than nanoindentation. This enables the determination of surface properties like stiffness and adhesion.

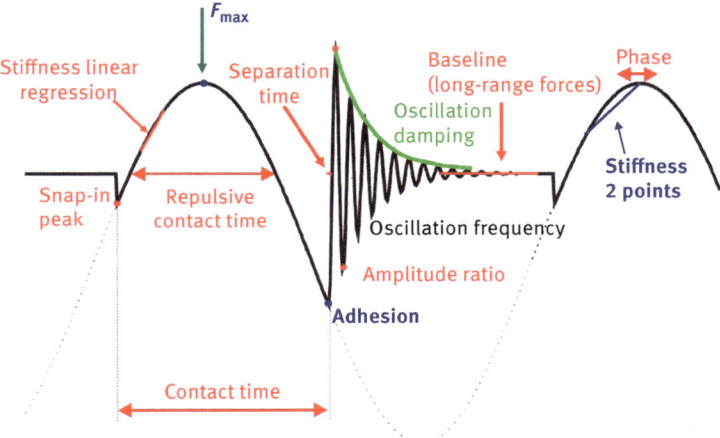

Fig. 3.31: Principle of pulsed force mode and relevant forces as well as material properties. Source: WITec GmbH.

Most AFM setups are limited in size and weight of samples as mentioned earlier, which limits their use for some components as glazing materials for PV applications. Fortunately, there are also some mobile AFM tools making it easier to analyze the surface properties of components or even complete PV modules without special sample preparation (Fig. 3.32, next page). This opens additional possibilities, even including potential on-site investigation of samples.

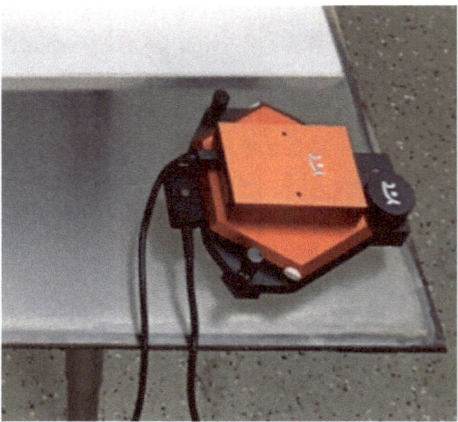

Fig. 3.32: Mobile AFM on solar glass plate with functional coating.

3.2.11 Contact angle goniometer

Elisabeth Klimm

The contact angle gonimeter measures the surface wettability via drop contour imaging. The term contact angle describes the angle of the three-phase boundary – liquid, gaseous, solid – between a liquid and a solid surface. A liquid is applied via syringe onto a surface in a defined quantity, within a defined time (Fig. 3.33). Applied could be a testing volume of, for example, 20 μL outgazed water at a dose rate of 5 μL/s in sessile modus. In order to place a static drop on a surface, the surface

Fig. 3.33: Syringe depositing a drop of water on the surface to be analyzed.

tension of the liquid must be greater than the free surface energy of the solid. The method of static drop called "sessile drop" is a standard method of contact angle measurement, where the tip of syringe is not in contact with the water drop. This method is usually used to analyze materials for PV applications. But of course there are numerous operation modes to analyze the wettability or surface tensions even by using different liquids.

For this purpose, the drop is illuminated on one side by a diffuse light source and the contour is monitored with a camera.

For solar glass, the water contact angle of its surfaces is measured (Fig. 3.34). Angles of 5° and larger are given as result, since angles below that cannot be resolved visually with the camera setup. The water then builds a film. For uncoated solar glass, the water contact angle of the applied drop is usually in the range of 30–60°, depending on the surface energy. Especially used is the drop contour for the analysis of functional coatings with water showing, for example, for super-hydrophilic coatings small angles below 5° forming a liquid film and for hydrophobic coatings drop contour angles around 80–90°, everything above 150° is called superhydrophobicity ("lotus effect").

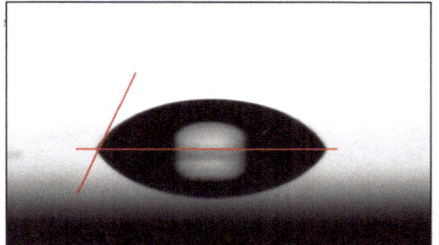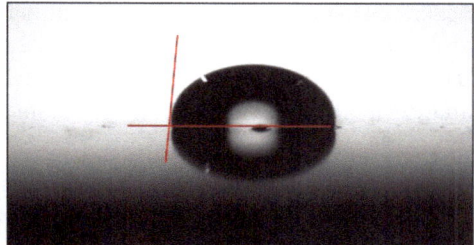

Fig. 3.34: Water droplets on solar glass with different surface properties and different contact angles.

In general, the measured result is highly influenced by the surface, and for reasonable results, a significant number of at least 5 measurements should be conducted. Deviations of up to 10° can be expected, since surfaces can be roughly structured and inhomogeneous. Typical devices can determine contact angles between 5° and 175° with a measurement inaccuracy of the video system of ±0.1°. The software determines the right and left contact angles and determines the average.

Another application for contact angle analysis is the measurement of flux for soldering of PV cell metallization, with a heated goniometer setup.

The roll of angle or dynamic contact angle is further used to characterize surfaces. It refers to the angle between the sample and the horizontal at which the water drop begins to move when the sample is lifted on one side. It can give hints on the adhesion forces of the surfaces. In dependence of the glass inclination the water drop starts to roll of, for example, for "smooth" solar glass surfaces at around 20°.

Changes in the contact angle of surfaces over time give a fast and easy indication if there is degradation in process, which cannot yet be seen with expensive optical devices and without high-resolution microscopic imaging.

3.2.12 Infrared imaging (IR) and Near infrared spectroscopy (NIR)

IR imaging uses dedicated camera systems to analyze the IR radiation of samples that are commonly used, for example, in thermal imaging cameras. In PV, there is a very established application to use IR imaging to identify hot modules in power plants since a higher temperature of a module can indicate issues of the module. This technology is used especially to look for hot spots in modules (see Section 4.2). For such applications to analyze full power plants, the IR imaging is usually performed from some distance to be able to screen many modules at once. Supporting technologies for this investigation are elevating platforms or in recent times more and more drones that can carry the camera.

Near infrared spectroscopy (NIR) is more dedicated to material analytics. The generated spectra can be used to identify the materials in the module (Fig. 3.36) and also to identify differences from the initial state and so also degradation effects. Since the IR spectrometers are typically handheld (see Fig. 3.35), they can also be used to analyze modules in the field for their state and bill of materials (BOM).

Fig. 3.35: Handheld IR imaging system for material analysis, in this case, backsheet analysis. Source: Gabriele Eder.

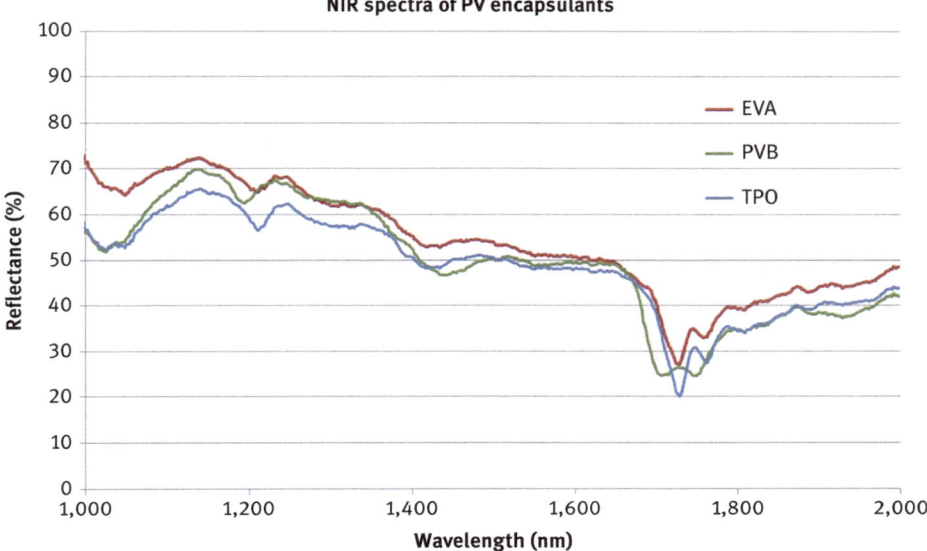

NIR spectra of PV encapsulants

Fig. 3.36: Spectra of different encapsulation materials taken with handheld IR camera through the glazing of a module. Source: Gabriele Eder.

3.2.13 Fluorescence imaging (FL)

Fluorescence imaging (FL) is a technology using fluorescence effects of materials. In PV, the fluorescent effects of polymers are analyzed, typically the measurements focus on the encapsulation material. EVA as the dominating material contains various additives such as oxidation stabilizers, UV absorbers, and cross-linking agents. Fluorescence is a form of luminescence, which is the emission of light by an activated material that has absorbed light or other electromagnetic radiation. The reemitted light has a longer wavelength than the absorbed radiation (e.g., UV light). Typical fluorophores (fluorinating materials) are degradation products of polymers and/or additives with chromophore/fluorophore groups. The fluorescence of materials can be extinguished when the fluorophores are destroyed by chemical processes, in PV for example, by "photobleaching" effects which lead to a decrease in fluorescence because of reaction processes with oxygen.

In PV, the typical setup for FL is containing a UV light source to illuminate the sample and a camera with a filter for the light of the illumination to monitor the fluorescent light (Fig. 3.37).

Since the fluorophores are mainly degradation products of the encapsulation EVA, FL can be used to analyze this component. This can be used to relate indoor and outdoor exposure or to compare the state of degradation of different products

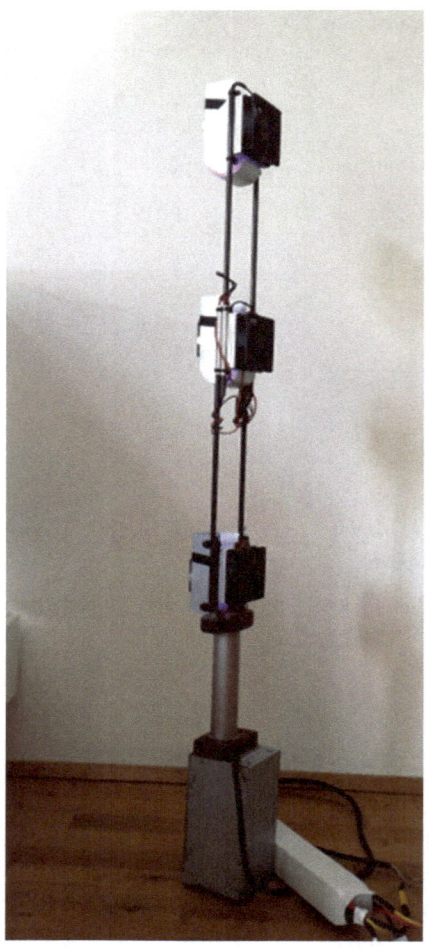

Fig. 3.37: Mobile fluorescence imaging setup with LED light source (380 nm), band-pass filter, and Li battery as power supply. Source: Gabriele Eder.

as long as it is ensured that the products consist of the identical materials. This effect can also be used to check if different modules consist of the same materials since fluorescence should be similar if the BOM is identical, which is unfortunately not always the case, as examples in Fig. 3.38 indicate.

Another application is to analyze secondary effects, for example, due to the bleaching effect of oxygen, which diffuses into the module from the rear side or from the edges (Fig. 3.39). This effect usually leads to different fluorescence intensities close to the edges of the cell, since the degradation products of EVA between cell and glazing in this area can be reached by oxygen from the rear side and is bleached. So FL can be used to check modules for cracked cells.

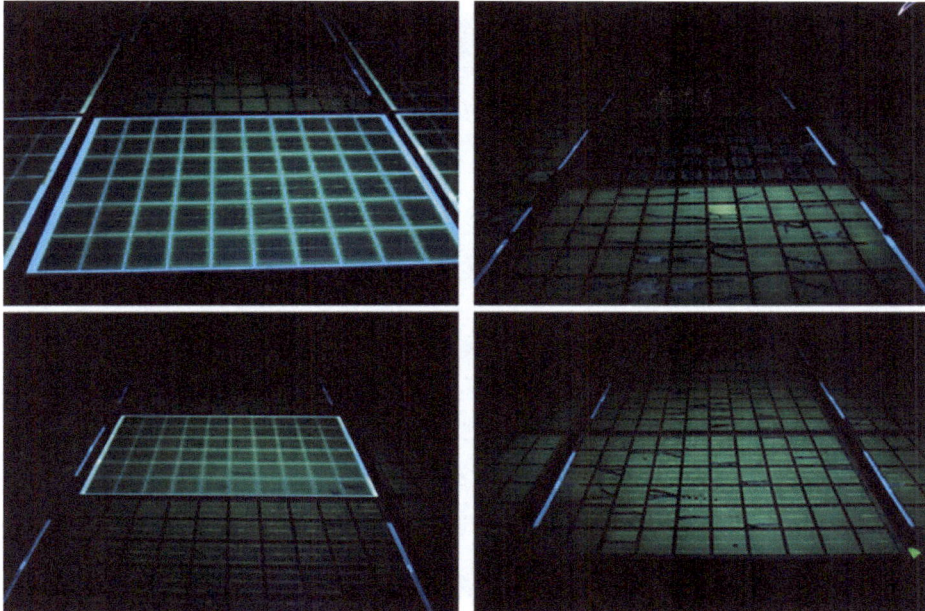

Fig. 3.38: Fluorescence image of a power plant with modules that were expected to have the same BOM. Source: Gabriele Eder.

Fig. 3.39: Fluorescence image of modules some time after a hail storm. The glazing of the modules is still intact but obviously some of the cells are cracked, enabling oxygen to reach the front layer of encapsulation material and bleach chromophores there. Source: Gabriele Eder.

Karl-Anders Weiß, Elisabeth Klimm

4 Loads for PV modules

During outdoor operation of technical products, the used materials are exposed to loads influencing their performance by causing degradation effects. This also applies for photovoltaic (PV) systems and modules. Since service lifetimes of 20, 25, or even more years are expected for PV modules, the investigation of degradation effects and relevant loads as well as stress factors is of special importance.

Loads can be categorized into external loads by weather and ambient conditions and internal loads, which can be related to the operation of the PV module/system itself. This chapter lists the most important stressors and load factors and describes their most relevant properties including typical influences and effects on materials or components with focus on the reliability theme.

4.1 External loads

External loads are a major influence factor on PV module degradation besides the materials, the design, and the processing. As visualized in Fig. 4.1, these loads are mechanical loads, temperature, radiation, moisture, salt (water), as well as chemical and biological or soiling loads. The influence of these factors may cause chemical and physical degradation processes on the inside of a PV module as well as on its outside, especially the surface.

4.1.1 Solar radiation and UV radiation

The electromagnetic radiation coming from the Sun is called, in PV terminology, global irradiation or solar irradiation G. Its spectrum and intensity depend on the geographical location, altitude, time, date, and weather at the specific place of measurement. The terminus air mass (AM) is used to describe spectral differences due to location and time. The spectrum of the irradiation outside the atmosphere is called AM 0, and the spectrum of irradiation after crossing the atmosphere and reaching the Earth's surface on the shortest way is called AM 1. In the PV industry, the spectrum at AM 1.5 is commonly used as reference spectrum for measurement of PV modules' performance and solar simulators since it describes the solar spectrum reaching the ground in Central Europe (exactly 48.2° North) around noon on a clear day (Fig. 4.2). With a peak intensity of 1,000 W/m^2, the AM 1.5 spectrum is described in the international standard IEC 904-3 (1989) part III.

https://doi.org/10.1515/9783110685558-004

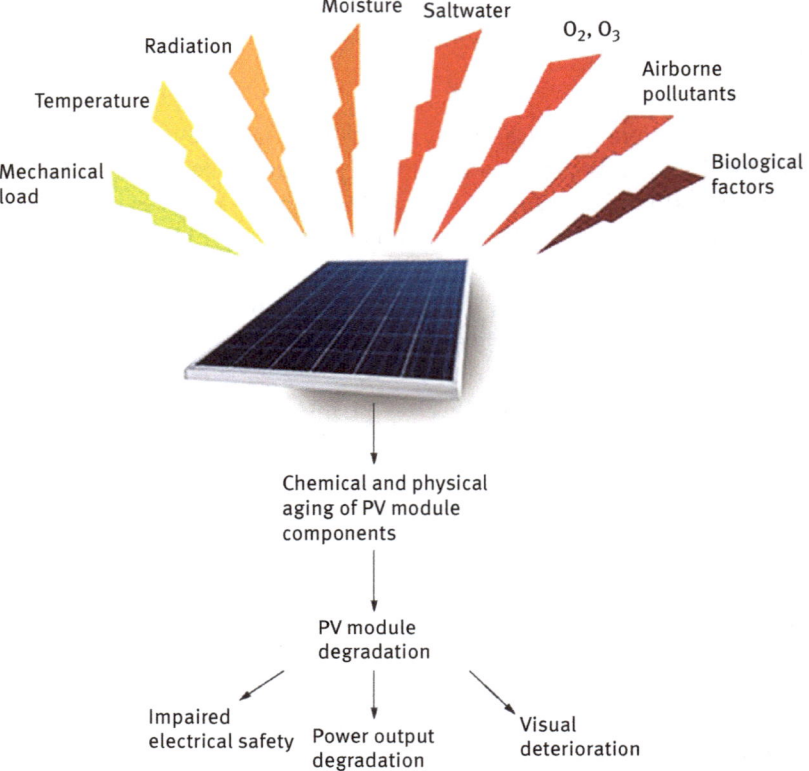

Fig. 4.1: External loads/stressors influencing PV module degradation.

Radiation causes different effects when it is absorbed by the material. On the one hand, there are positive effects like the generation of electric power in PV cells, but on the other hand, there are also numerous negative effects causing degradation. The absorbed radiation increases the temperature of the material leading to temperature-dependent effects but it can also directly cause damage. Each molecular bond has a characteristic energy which is necessary to break it. Thus, photons with an energy which is at least as high as this dissociation energy E_D can destroy the chemical bonds and cause changes in the material and its properties.

Especially the chemical bonds in the polymeric materials in PV modules are susceptible to certain ranges of ultraviolet (UV) irradiation, and it is very important to understand the spectral sensitivity of the materials. The energy of the photon is of great importance regarding the radiation's impact on the polymer. The bonding energies and the necessary wavelengths to crack some of the most relevant bondings in polymeric materials are listed in Tab. 4.1. If sufficient energy in the form of radiation or heat is supplied, an initial chain scission step can occur, followed by

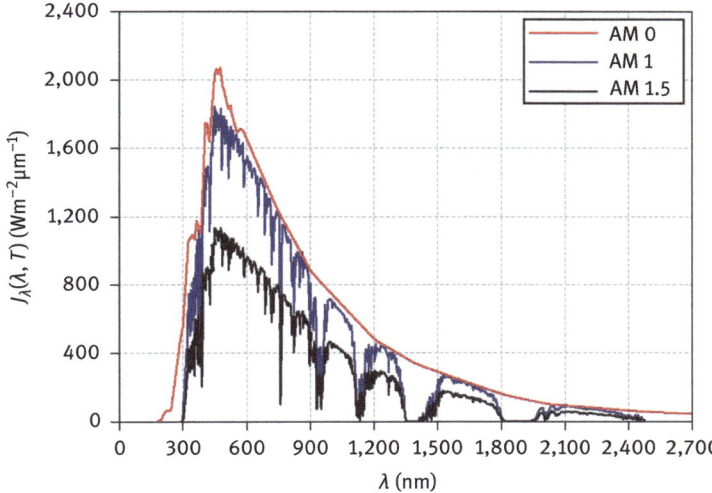

Fig. 4.2: AM 0, AM 1, AM 1.5: solar spectra outside the atmosphere (AM 0), at vertical incidence at the equator (AM 1) and reverence spectrum for solar application with zenith angle of 48.2° (AM 1.5).

the formation of two alkyl radicals. Without the necessity of any further energy input, those radicals may react with molecular oxygen to form a peroxide radical. In the case of additional energy input, the peroxy radical can abstract a hydrogen atom from an alkyl chain, resulting in a hydroperoxide and an alkyl radical which both can react with oxygen to form a peroxide and a peroxy radical, respectively. This way, hydroperoxides are generated, which may decompose to an alkoxy and a hydroxyl radical. Those radicals may break further C–H bonds under the formation of alcohol or water and alkyl radicals. The latter ones in turn react with oxygen to peroxy radicals.

Tab. 4.1: Binding energies of some primary valence bondings and the respective wavelengths.

Bonding	Bonding energy (kJ/mol)	Wavelength (nm)
C=C	837	143
C=O	729	164
C=S	540	222
C–C (aromatic)	519	231
O–H	460	260
C–O	364	329
O–O (peroxides)	268	447

The energy that is necessary to destroy the polymeric chains usually requires photons of the UV part of the solar spectrum. Therefore, this part of the solar spectrum is of specific relevance when dealing with reliability issues of PV. As parts of light's longer wavelength influence the temperature of materials and the measurement of UV irradiation is neither cheap nor (usually) precise, usually the intensity and dose of the full solar spectrum are measured.

For the calculation of the UV dose, a rule of thumb is the estimation of 5% UV share of the full-intensity global irradiation. This approximation can be utilized for the estimation of the UV loads but there are significant differences between locations, and the UV share of the incident light varies depending on the season. Examples of measured UV/G ratios for a test site in Freiburg, Germany, are shown in Fig. 4.3 and the differences between test sites and the seasons in Tab. 4.2.

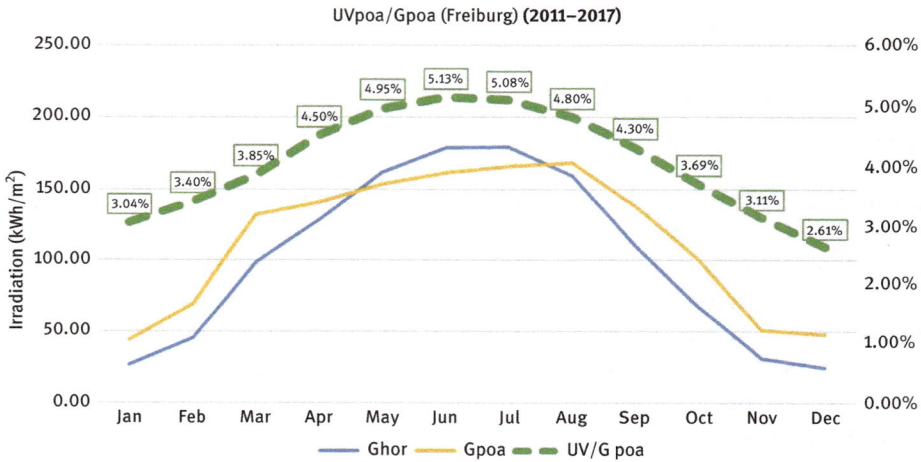

Fig. 4.3: Measured ratio of UV/G in plane of array (POA), 45° looking South in Freiburg, Germany.

Tab. 4.2: UV/G ratio of different tests sites. Values are measured in POA. The test sites are Freiburg, Germany (FRBG); Mountain Zugspitze, Germany (UFS); Gran Canaria, Spain (GC); and Negev Desert, Israel (NEG).

	Latitude (North)	POA$_{Tilt}$ (to the south)	Altitude (m)	Sum of G (POA) (kWh/m²a)	Sum of UV (POA) (kWh/m²a)	UV/G	UV/G winter–summer
FRBG	48°	45°	265	1,373	60.01	4.4%	2.6–5.1%
UFS	47°	45°	2,650	1,702	81.68	4.8%	2.9–5.8%
GC	28°	23°	5	2,274	100.47	4.4%	3.7–5.1%
NEG	31°	31°	300	2,401	95.08	4.0%	3.4–4.4%

4.1.2 Temperature

Temperature has a large impact on most chemical processes, and is thus very significant with regard to degradation processes. The influence of temperature on the speed of reactions, as accelerating factor, can be described by different mathematical models. The most common one is the Arrhenius equation (eq. (4.1)):

$$k = A \cdot e^{\frac{-E_A}{R \cdot T}} \qquad (4.1)$$

where A is the material factor, E_A is the activation energy $(J \cdot mol^{-1})$, R is 8,314 $J \cdot K^{-1} \cdot mol^{-1}$ – universal gas constant, T is the absolute temperature (K), k is the reaction rate.

The acceleration effect of temperature depends on the specific reaction and on the material. This leads to different acceleration factors for one product when multiple reactions take place under changing/not identical conditions, for example, at different temperature ranges or in dry or humid conditions. For correct description of degradation effects with mathematical models (see also Chapter 9), it is important, so first select an appropriate model and then the relevant set of parameters.

The temperature can be determined via reliable and fast measurements with cheap sensors such as thermoelements or Pt-100 resistivity sensors. As temperature data are available for many locations, real-life data can be used as a basis for aging tests. The heating effect of solar radiation on materials and sensors has to be taken into account. Thus, significant differences between ambient temperature (macroclimate) and sample temperature (microclimate) have to be expected. The microclimate at a sample should be regarded as the relevant one for degradation effects, since differences between ambient and module temperatures can be as high as 30 K as shown in Fig. 4.4. In addition, the issue of determining surface temperatures under irradiation has to be dealt with. Also often found mistakes in data are due to an inadequate measurement setup leading to wrong temperature data. Thus, temperature sensors should be selected and placed at the right sample's position with great care, for example, checked by a second sensor.

Besides the acceleration of chemical reactions, mass transport reactions are also accelerated at high temperatures. At higher temperatures, water vapor or oxygen from the atmosphere is more likely to ingress into the module causing hydrolytic or oxidative reactions. At the same time, positive effects such as easier outgazing of degradation by-products, which reduces the corrosive stress within the module laminate, can occur [34].

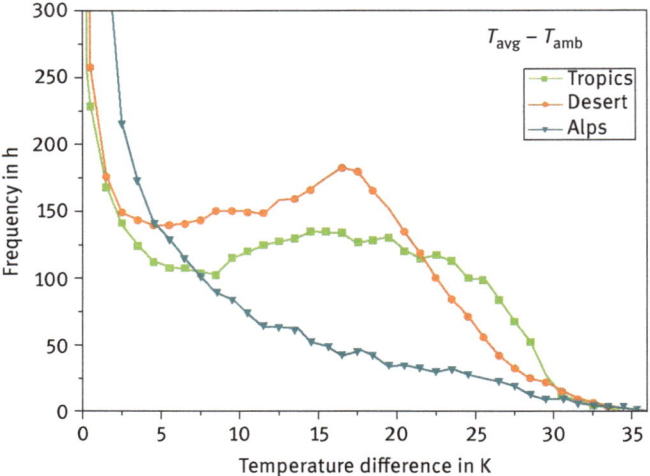

Fig. 4.4: The module temperature can vary dramatically from ambient temperature, depending on the climatic conditions and the absorption properties of the PV module [35].

4.1.3 Humidity

A certain amount of humidity is available in all climates at all times. Hence, moisture has to be taken into account when it comes to exposure of PV modules, especially since its influence on degradation effects is very dependent on the materials used.

While, for coatings, often the condensation of humidity on the outer layers due to dewing is of major relevance, polymeric degradation effects like hydrolytic degradation processes occur inside of a module. Consequently, the meaningful moisture level has to be determined with respect to the relevant effect of humidity and the site of investigation. While the relevant measurement of the relative ambient humidity is external and easy, what requires only simple sensors, the relevant value for the degradation of samples, is the intrinsic humidity level directly at or in the material bulk (microclimate). The direct use of ambient climatic data for estimation of the bulk samples' humidity level is usually not suitable. The necessary direct measurement on the surfaces or in the material itself is often not possible or also influenced by artifacts, for example, due to the humidity capacities of the sensors themselves and the low absolute humidity levels in encapsulation materials. Usually the best way is the calculation of the microclimatic humidity based on ambient humidity and material temperatures. For these calculations, the temperature is of major relevance since the humidity capacity of most materials strongly depends on the temperature, especially the humidity capacity of air. Therefore for objects with a higher surface temperature as the ambience, like PV modules in operation, the relative humidity is much lower than in the measured ambience. On surfaces of objects which are colder than the ambience, condensation can occur. One example is PV modules during clear nights, especially when humidity increases significantly (e.g., >90%), which can cause condensation.

On material level (bulk and surface), a definition of the critical sample humidity level, above which the humidity is relevant for degradation processes or above which the sample can be seen as humid or wet, is often also not easy or not possible. Therefore, one approach is the use of values, such as the **"time of wetness (TOW)."** TOW is used to quantify times above a certain level of humidity and for a comparison of locations. Times with relative humidity >80% are often as rule of thumb counted as TOW. Material research determines the critical humidity level for a specific material; hence, the critical level for a product strongly depends on the used materials even at the same location.

Exterior PV module components are usually chosen to be humidity stable. But atmospheric humidity still impairs the interior PV module components. The **water vapor diffusion (WVD)** proceeds from the ambience into the module. Ways of ingress are through the back-sheet foil and the edges, for glass–glass modules only through the edges. Diffusion proceeds in the encapsulation layer reaching finally the sensitive live parts, the intercell connectors, the cell metallization, and the solar cell. Those metallic and electrical live components can subsequently undergo corrosion reactions.

Also hydrolysis of the encapsulation material or the backsheet core layer can lead to an impaired PV module performance or electrical safety. Thus, the full **permeation process**, which is active in a PV laminate including the described diffusion process, is crucial for the understanding the humidity-induced PV module degradation. Permeation is a mass transport process of molecules through a membrane from an exterior to an interior environment, which is accomplished by diffusive processes. Chemical reactions and diffusive processes are accelerated by increasing temperatures, in most cases calculated with the Arrhenius equation, which has been described in Section 4.1.2, eq. (4.1). The influence of this temperature dependency of the water uptake in the encapsulation material in a PV module is displayed in Fig. 4.5.

The water ingress through the encapsulation layer, reaching finally the center point above the solar cell was simulated over a period of 20 years. Plotted are the resulting moisture levels for one module design with two different backsheet materials, which are PET and PA, at four climate zones: tropic, alpine, arid, and moderate. The results show significant variations in the time-dependent water uptake, mainly due to the differences in relative humidity levels and ambient temperature at the different exposition sites, but also because of the different permeability of the backsheet materials. The duration of the initial accumulation phase, before the equilibrium state is reached, is found to be highly dependent on the ambient temperature. For cold climates, such as in the alpine zone, the accumulation time takes about 10 years, while it only takes 2–3 years in hot, arid climates. In contrast, the actual amount of

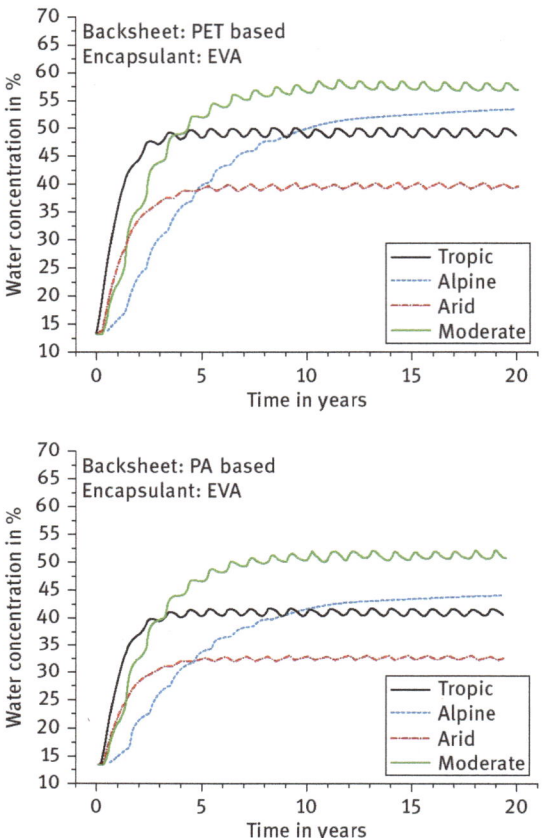

Fig. 4.5: The simulated water ingress into a PV module for two different backsheet materials (PET and PA) and four different climates (tropic, alpine, arid, and moderate) displays significant differences in the time-dependent water ingress, which is influenced by the relative humidity, the average temperature, and the permeability of the backsheet [36].

water present inside the PV module is mainly determined by the humidity of the surrounding atmosphere. Therefore, water concentrations in our examples of humid climates (tropic, moderate) are up to 100% higher than in hot and dry climates (arid). The reason for the counterintuitive relationship between the water content in moderate and tropic climates is most likely the outgassing of previously incorporated water which proceeds more efficiently in climates with high daytime temperatures. The water uptake is also influenced by the permeation properties of the backsheet, caused by different temperature dependencies of different backsheet materials, especially the water vapor transmission rate (WVTR), as shown in Fig. 4.6.

These different temperature dependencies of the **water vapor transmission rate** result in differences in the acceleration of both water vapor uptake into the module

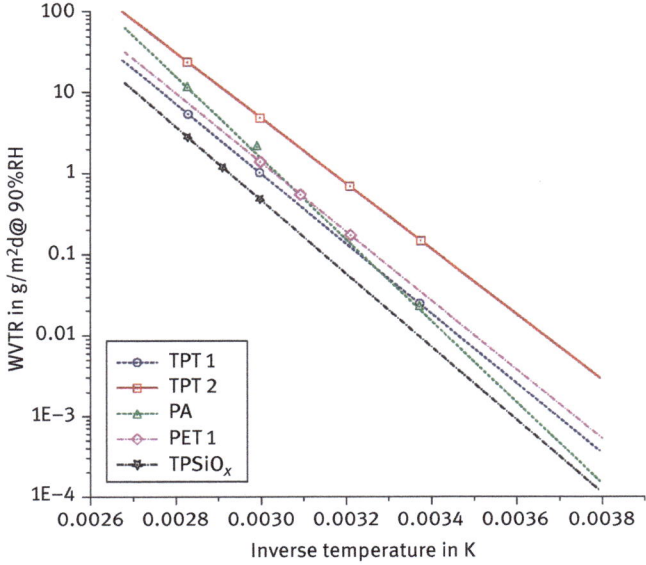

Fig. 4.6: The measured water vapor transmission rates (WVTR) of several backsheet materials display significant differences with regard to their temperature dependence [36].

and the outgassing of water vapor. Therefore, for example, for a module with a PA backsheet, less water vapor can ingress into a module through the backsheet (PA) during cold, for example night, periods. In addition, more water can outgas during periods of high module temperatures than with PET-based Tedlar–PET–Tedlar back-sheet. It is because during the daytime the module is hotter than the environment due to solar irradiation.

4.1.4 Mechanical loads

The majority of natural mechanical loads on PV modules are caused by
- wind,
- rain,
- snow,
- hail,
- ice, and
- abrasion.

Unfortunately, also man-made mechanical loads have to be taken into account for mechanically damaged modules. A trained eye is needed to spot these damages in analysis or when degradation effects, which are to be explained by mechanical loads, are investigated. Those mechanical loads induced by people walking over modules or tools falling on modules can cause significant damage, and therefore these impacts should be strictly avoided. Since these impacts are highly nondeterministic and not possible to be predicted with mathematical and physical methods, they are not included here but should anyway be strictly avoided by training and surveillance measures.

The **wind load** impacting the PV module derives from the pressure distribution around. Besides a zone of excess pressure in front of the module and of negative pressure behind the module, turbulences are induced and as a result, a displacement of the module perpendicular to the plane of mounting is caused. This displacement results in a mechanical stress, which is most harmful for the cell interconnectors and the cells and may therefore lead to material fatigue and the formation of cell cracks.

Of the listed **precipitation impacts** such as rain, hail, snow, and ice, hail is the only one which has an instantaneous impact on the module. In dependence of the respective hail grain size, hail may lead to glass breakage and in turn to an impaired electrical safety and stability of the module. This is mainly an issue if no tempered glass is used as glazing. While rain droplets can induce vibration of the PV module, snow and ice accumulating on the surface of a PV module can cause high static mechanical loads of up to several thousand pascals. It may lead to delamination, glass, and cell breakage, or the detachment of the frame, which is surprisingly the most common effect of heavy snow load damage.

An additional mechanical impact, which is mainly relevant for exterior materials like back- and front-side materials, like functional coatings, glazing, or backsheet, is abrasion. **Abrasive loads** are of special relevance in areas with high wind loads and dry conditions, leading to sand storms or in general a higher particle concentration in the air. If this is the case for a specific site, it makes sense to have a look on the typical wind speed and direction around the year and choose appropriate materials, for example, more rigid or thicker backsheets or specific coatings. The example given in Fig. 4.7 shows data of a site with very stable main wind direction.

Abrasive loads related to soiling (compare Section 4.1.7) can also be caused by **module cleaning** measures. There are dry and wet cleaning methods and also cleaning by hand with brushes as well as by robots or other machinery. These measures to remove sand and soil from the surface by nature cause abrasive loads since the particles are moved along the surface. Many manufacturers, therefore, exclude degradation due to cleaning from their warranties or clearly describe which cleaning methods are to be applied. It is recommended to include the expected soiling effects and the related cleaning in the load estimation and also the reliability testing before installation.

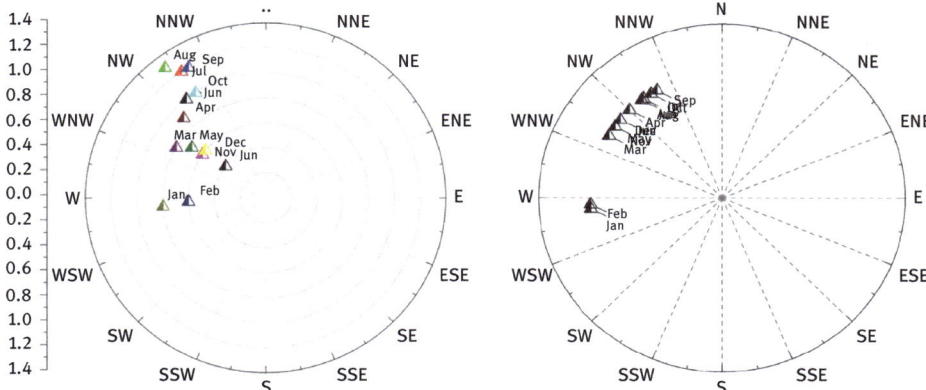

Fig. 4.7: Distribution of a measured main wind speed and direction (left) and direction (right) for 1 year (2015) at a test site in the Negev desert, Israel.

4.1.5 Corrosivity

Elisabeth Klimm

Besides polymer degradation, the degradation of the inorganic components in a PV module is one of the most important aspects of PV module degradation [37]. Significant decrease in PV module performance is caused by the corrosion of the cell, that is, the SiN_x anti-reflective (AR) coating or the corrosion of the metallization, that is, solder bonds and the silver (Ag) fingers [6, 38–40].

As stated earlier, metal corrosion is an electrochemical process, which is the reason for the necessity of an aqueous electron conduction-enabling environment to be present. Corrosion is a surface process in which atoms at the surface of the metal enter the solution as ions, and at the same time, electrons migrate through the metal to a site where they are consumed by species in contact with the metal [41]. The metal ions may also combine with other species in solution, and compounds such as oxides or hydroxides can precipitate.

A corrosion process takes place spontaneously when the change of the free enthalpy $\Delta G°$ during this process is <0. For an electrochemical reaction, ΔG is proportional to z, which is the number of electrons transferred in the reaction as well as to the difference $\Delta E°$ between the standard potentials $\Delta E°_{red}$ and $\Delta E°_{ox}$:

$$\Delta G° = -zF\Delta E°$$
(4.2)

The environmental conditions can be integrated in this relationship via the Nernst equation:

$$E_{red} = E°_{red} - \frac{RT}{zF} \ln \frac{a_{red}}{a_{ox}}$$
(4.3)

where E_{red} is the half-cell reduction potential at the temperature of interest, E_{red}° is the standard half-cell reduction potential, R is the universal gas constant, T is the temperature, a is the activity for the relevant species, and F is the Faraday constant. In a PV module, the front-side cell metallization, that is, the grid that mainly consists of silver as well as the Sn/Pb-capped copper intercell connectors and the rear side metallization, which is typically made from aluminum, are the corrosion-susceptible components. An electrolytic environment is created when water permeates into the module, that is, into the encapsulation creating an ion conduction environment. The corrosion kinetics is either determined by the through-passage of the ion or by the diffusion of the electrons. Besides the silver grid on the silicon cell, the intercell connectors can undergo corrosion reactions.

But how can the environmental load be classified? The atmospheric corrosivity is a location-dependent specific load factor and so the relevant value for the PV system and material reliability.

The complete process of a corrosivity assessment is starting with the measurement on the defined test site, continuing with the analysis of exposed samples, and ending with the categorization and ranking of the test sites atmospheric corrosivity.

The atmospheric corrosivity assessment, is carried out by the exposure of standard specimens for 1 year to the atmosphere at the respective location according to the international standard ISO 9226. The standard specimens are flat plates of four standardized metals: aluminum (Al), copper (Cu), steel (Fe), and zinc (Zn). The corrosivity of the exposure locations is deduced from the corrosion rate, calculated from the loss of mass per unit area of these standard specimens after the exposure period.

The quality of the corrosion coupons from the metals of unalloyed carbon steel (Fe), zinc (98.5% Zn at min.), aluminum (99.5% Al at min.), and copper (99.5% Cu at min.) has to be in accordance with the ISO 9226 standard. The samples have to be exposed to the atmosphere without shelter on a rack facing south (in the northern hemisphere) or facing north (in the southern hemisphere). The standardized exposure of the setup with the coupons is in horizontal position and with an inclination angle of 45° in every location. The corrosion setup must be exposed with a minimum height of 0.5 m above the ground to avoid contact with the surrounding vegetation. Samples are not to be touched or cleaned during the exposure period (Fig. 4.8).

The corrosion rate of the samples is determined according to the standard ISO 8407 after 1 year of exposure (Fig. 4.9) – as a difference between the original weight and the final weight. The corrosion rate r_{corr} derived from the mass loss measurement in grams per square meter and year is calculated for each sample according to ISO 9226.

The values obtained from the corrosion rates are used as classification criteria for the evaluation of atmospheric corrosivity according to the standard ISO 9223. The atmosphere at the test site is classified from C1 to CX according to the corrosivity for each metal separately.

Fig. 4.8: Metal coupons for determination of atmospheric corrosivity before exposure.

Fig. 4.9: Metal coupons for determination of atmospheric corrosivity after 1 year of exposure on test site at Gran Canaria, Spain, with very corrosive climate.

4.1.6 Other chemical loads

The presence of oxygen is crucial for all oxidation reactions, which constitute the most important polymer degradation mechanism. Similar to humidity, oxygen has to permeate through the backsheet material in order to initiate oxidation reactions in the encapsulation material. The **oxygen transmission** properties of different backsheet materials differ extremely for the different PV materials. The measured oxygen transmission rates of several different materials can be found in Fig. 4.10.

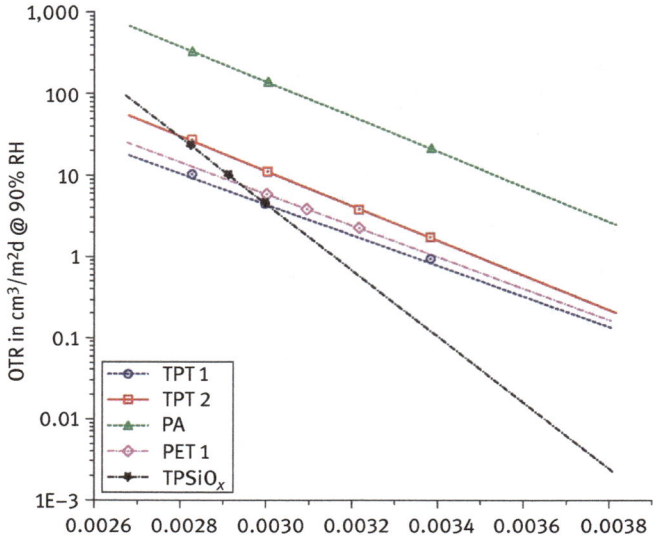

Fig. 4.10: Measured oxygen transmission rates (OTR) of several different materials [40].

The presence of oxygen limits all oxidative reactions, and therefore a high oxygen permeation rate can result in an acceleration of the polymer oxidation reactions, which can result in chain scission and cross-linking reactions.

In addition to oxygen, other gases, which are present in the atmosphere or which can be formed under certain conditions, can ingress into PV module. Then these **atmospheric gases** cause degradation reactions or lead to a degradation of the external PV module components such as the frame. Gases such as NO_x, SO_x, and O_3 can be washed out by rain events and precipitate on the surface. In the presence of water, nitric acid and nitrous are formed from NO_x, and sulfuric acid as well as SO_3 come from SO_x. Those acids can cause hydrolytic bond cleavage in certain polymers like polyamide. Also under the impact of dew and fog, minimal pH values of 1.5 were measured in Germany. The influence of corrosive gases on the PV module degradation has been to date hardly investigated. High levels of chloride deposition can be found in coastal regions, but the chloride content in the air decreases rapidly with increasing distance to the shoreline. **Chloride ions** can move through the layer of corrosion products and arrive at metal surfaces where they accelerate the corrosion process [42]. For the anions in this process, a catalyst role can be considered [43]. **Sulfur dioxide** (SO_2) is mainly released as a side product from the combustion process of oil or coal due to a certain sulfur content. Therefore, main sources of SO_2 pollution are energy plants and car exhaust fumes. Thus, SO_2 levels are increased in industrial environments. In the past two decades, luckily the usage of low-sulfur fuels resulted in a drastic decrease of SO_2 in the atmosphere. The solution of SO_2 in water may result in the formation of sulfuric acid, which accelerates metal corrosion. The SO_2 content is determined via the SO_2 deposition rate or per concentration measurements. Figure 4.11 shows the distribution of atmospheric corrosivity on the global scale as map. It was created using a geographical information system [44]. Thus, the chloride deposition is only important close to the sea, literature is ambiguous, some papers say in areas less than 500 m away from the sea [45]. Since sea salt, that is, NaCl, is a strong electrolyte when diluted in water, reactions requiring conduction as well as ion exchange on the surface of metals are accelerated under the presence of Cl⁻. The amount of chloride in the atmosphere is determined by the chloride deposition rate, which can be measured either by the "wet candle" method or by conductivity.

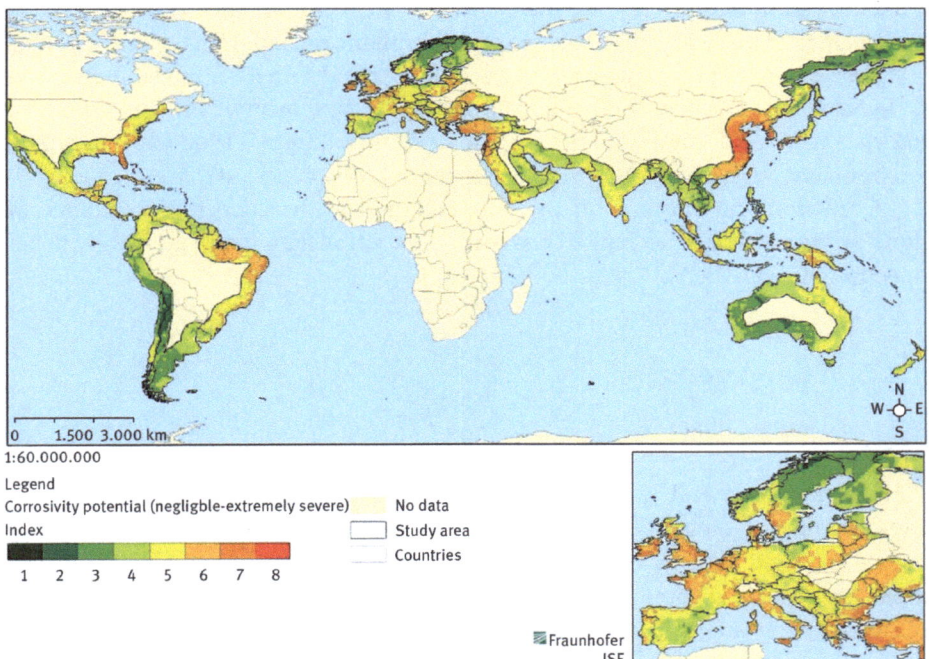

Fig. 4.11: The distribution of atmospheric corrosivity on the global scale was created using ArcGIS considering the chemical factors such as salinity, SO_2 pollution, and relative humidity of the atmosphere [44].

Atmospheric corrosion in coastal regions can e.g. be classified into eight color codes (Fig. 4.11). The classification is based on the three measured key corrosion factors: relative humidity, airborne salinity, and SO_2 pollution. Hence, a corrosivity map with a color code ranging from green (very mild atmospheric corrosion) to dark red (extremely high corrosion rate) can be generated, showing that the atmospheric corrosivity is very heterogeneously distributed around the world. In some regions, such as China, Brazil, the Arabian Peninsula, or the east coast of the United States, high atmospheric corrosivity levels can be found. Their high corrosion level can be attributed either to very high airborne salinity or to the interaction of the factors humidity, salinity, and air pollution. Especially eastern China with its high atmospheric corrosivity is a good example for this synergetic effect [44].

The degradation of metals by airborne pollutants was attributed to the deterioration of SO_2 and in special environments also to chlorides. Therefore, those pollutants receive attention in atmospheric corrosion research. Environmental protection laws enacted about two decades ago in industrialized countries led to a notable decline in the SO_2 levels. However, At the same time, concentration levels of other air pollutants, such as the nitrogen oxide, and secondary pollutants, such as ozone, remained rather constant or even increased slightly. This has increased the relative significance of these air pollutants in the atmospheric degradation of materials [46].

Assessment of other chemical loads could be done with passive samplers, especially sampler for nitrogen dioxide, sulfur dioxide, and ozone. Low price samplers are vials with metal meshes to trap and absorb the pollutants. Their exposure in the environment takes up to 30 days, in dependence of the expected load.

Extraction and analysis for the determination of the collected pollutants are done by the producers accredited analytical laboratory with their extensively developed in-house methods.

4.1.7 Soiling loads

Elisabeth Klimm

Particle deposition and accumulation of dust and soil or sand on PV modules can cause significant losses of the system's performance. This soiling of glass surface can reduce the yield even by up to 66% within a few months (Fig. 4.12). Although this topic has recently become increasingly relevant, as shown by the increasing number of current publications, further research has to be carried out, taking site-specific impacts into account in order to provide suitable mitigation approaches. Therefore, coatings that achieve a higher transmission on an annual average due to their dirt-repellent character seem to be reasonable. Soiling itself is not an issue for reliability since the effect is – according to the definition – not permanent, but it comes along with and causes several issues related to reliability.

Soiling losses are relevant only for very specific regions, particularly in regions of the sun belt, around the 30th latitude, with an annual solar radiation of up to 2,200 kWh/m^2, being almost twice as high as in Germany. However, the dust and sand pollution in these desert regions is high and the related losses due to soiling can

Fig. 4.12: Soiling on PV modules at the outdoor exposure test site of Fraunhofer ISE at Gran Canaria, Spain, 2010. Brown soiled surfaces compared to blue cleaned surface.

be significant, as shown in Fig. 4.13. Particles clog the surfaces of the solar systems, mainly as a result of extreme weather events and deposition, absorb and scatter sunlight, and thus reduce their efficiency. For this reason, it is important that the surfaces of the solar systems remain free of deposits. A necessity of more frequent cleaning has the disadvantages of additional impacts (and so potential degradation) to the surfaces, cost, and consumption of clean water, which is a precious commodity especially in the sunny regions.

In order to understand the transportation processes of dust and soil, as well as their accumulation onto surfaces, the chemical and physical properties are still under research. Some testing laboratories claim to have found the one and only – suitable for all locations – test dust. But just as with the geographical climate zones, there are many different influencing factors on the properties and composition of the dust; it sums up to the fact that there are different effects found when testing alike surfaces at controlled conditions with different dust types (Fig. 4.14). This leads again to the phrase: "location, location, location!" (Larry Kazmerski, IEEE PVSC 2018)

In soiling simulation in lab scale, by having different types of dust applied in wet or in dry matter deposited in a homogenous layer onto the surface by using, for example, the self-developed device using air pressure, further clogging patterns can be investigated. Yet, as soon as artificially weathering loads such as condensation are applied to the procedure, similar clogging patterns are not found so far. They are still under investigation.

In general, there are two stages of accumulation. First, specific fine particles adhere in a homogeneous layer directly to the surface, which cannot be determined at direct sight. Still it can be seen looking at the surface in a very steep angle, or it can be determined by the performance loss of the system. The significant reduction of performance is caused by losses of up to 10% in transmittance of the glazing. At the second stage, more particles bond to the first particle layer. In combination with

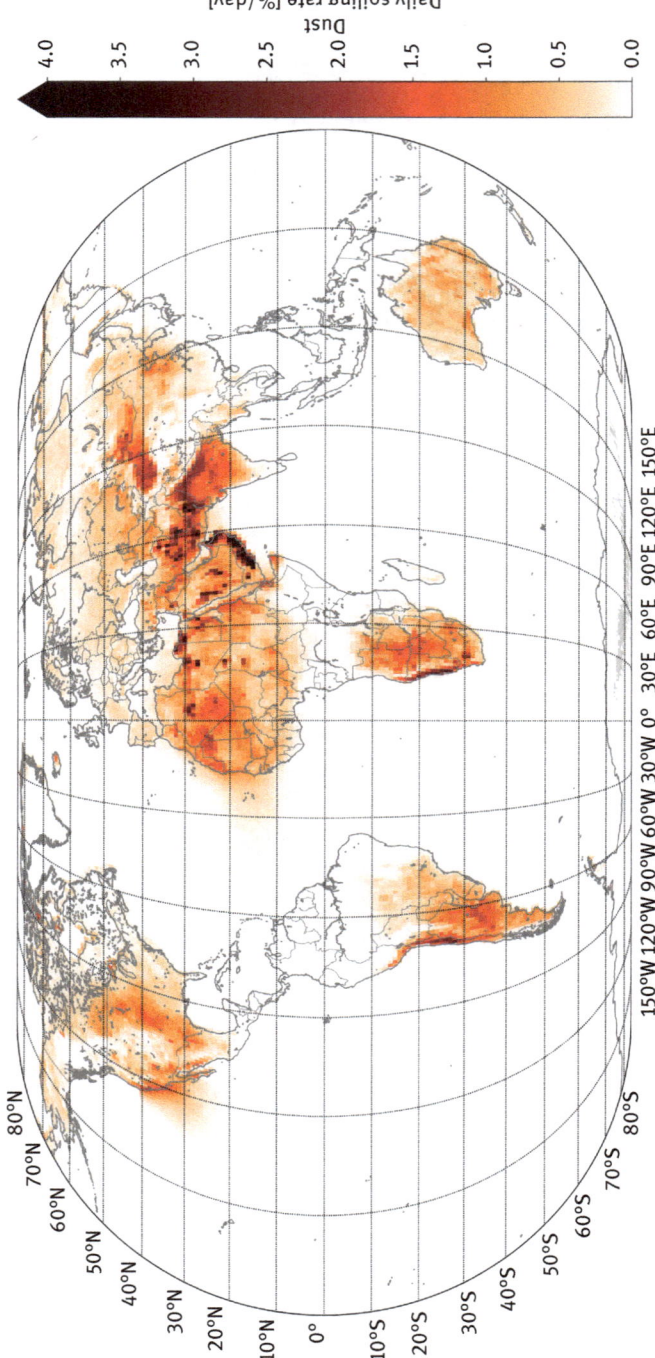

Fig. 4.13: Estimated daily soiling rate map showing the global distribution of the soiling risk. Source: Julián Ascencio-Vásquez, LPVO-UL.

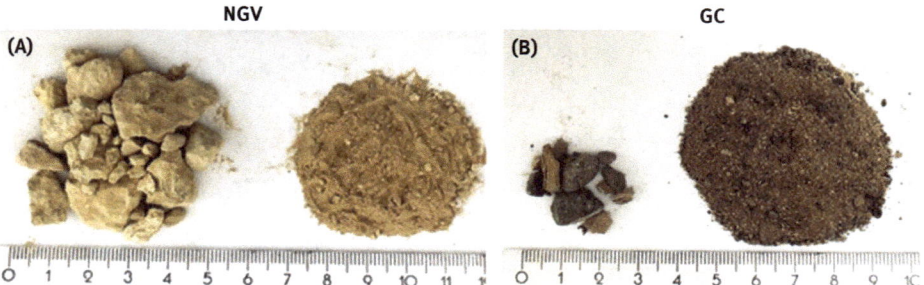

Fig. 4.14: Different types of dust from outdoor test sites in the Negev desert (A) and on the Canarian Islands (B).

weathering loads the bonds to surface and other particles can become very strong. This effect is often called cementation. In general, these bonds can still be broken up, and the surface can be cleaned. The soiling effect is reversible. Tests in Germany under outdoor conditions indicate that even with a significant dust layer, natural weathering effects are able to clean the surfaces all-season. For the weeks of exposure, no degradation, cementation, or likewise is found, and the surfaces were cleaned by the occurring rain and recovered to their initial transmittance, being directly related to performance. Irreversible degradation of surfaces in relation to soiling occur, for example, by abrasion (see Section 4.1.4) due to cleaning or because of heavy weather events such as dust storms with large particles impinging the surface. Another, in literature described, nonreversible degradation effect is the chemical alteration of surface properties by particles bonding to it. Of interest is the fact that a large quantity of dust deposited at once onto a surface has a slightly lower adhesion to the surface. In tests, a thick dry dust layer could easily slide off as one body already at an inclination angle of 30°.

The performance assessment of the soiling loads can be done with the help of transmittance measurement via Fourier-transformation infrared spectroscopy. Glass surfaces are optically measured before and after soiling. In parallel, the deposited amount of dust is weighed. Dust is very efficient in reduction of transmission. Figure 4.15 shows the results for a deposition of only 1 g/m² of dust.

The way of measuring soiling load outdoors can either be done by looking at the PV current I_{SC} of PV modules or setting up a soiling sensor. Meanwhile, there are market available sensors, but it is advised to ask the question how meaningful the results are, when the surface of the sensor is different than the surfaces found in the PV module field. A very easy do-it-yourself setup is a single-cell PV module and a chosen glass or polymeric glazing added on top of the cell, which is representative for the PV module glazing material. Monitoring the mini-modules' power output and the optional backside temperature is recommended. The sensor should be in plane of array with the modules and at the same height. It could be of use to install

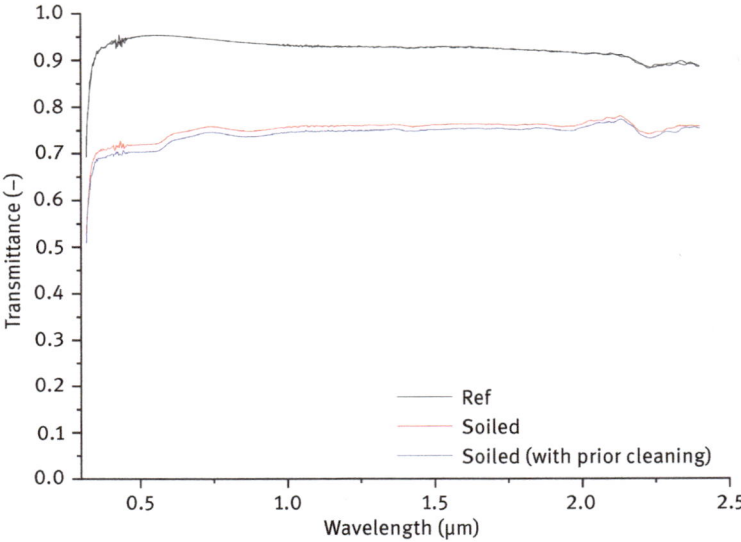

Fig. 4.15: Transmittance of solar glass in initial state and after artificial soiling with only 1 g/m² of dust. Prior cleaning of sample does not have significant effects.

an additional sensor at the bottom line of the modules to also monitor the full effect of the ground and vegetation, which can increase the soiling load. Usually a stronger soil layer is also accumulated at the bottom line of a PV module, which can – in dependence of the PV module setup – also shade parts of the cells and so even lead to hot spots (Fig. 4.16).

Fig. 4.16: Dust layer at the bottom of PV module with frame.

4.1.8 Electrical loads

External electrical loads impacting on PV modules are typically only arching events. These impacts are so severe that they often directly lead to a destruction of the module, for example, by broken front glass, melted connecting cables or junction box, burned holes in the backsheets, or other major impacts, which immediately ends the function of the module and usually the complete system. The events and the effects are by nature very spontaneous and can therefore not be described as or be handled like degradation effects.

4.2 Internal loads

Some loads that can cause degradation or affect reliability are also caused by the module itself and its components or its operation.

4.2.1 Chemical loads

Chemical loads within the module are caused by materials that do have an effect on other materials or by degradation products. Especially aggressive substances can be an issue for such effects. It is well known that the degradation of ethylene-vinyl acetate (EVA), the most common encapsulation material, leads to the formation of acetic acid. Acids chemically attack metals and corrosion of metallic components or materials within the module can be caused if the design does not allow the acetic acid to leave the module. In the beginning of the PV module industry, aluminum cell connectors have been used in glass–glass modules. Since the acid could not exit the very tight glass–glass design, severe corrosion of the connectors was induced leading to connector breakage. It led to the loss of functionality of the modules. Some degradation effects of the cell metallization have also been linked to the attack of acetic acid.

Fluxing agents, which are used to optimize soldering, are developed to affect the surface conditions of the soldering substrate, in case of PV modules the surface of PV cells, and so can influence the adhesion strength of the encapsulant to the cell. There are delamination effects in modules reported, which are linked to an excessive application of the flux agent. The reason was that the agent also covered larger areas of the cell surface, leading to bad adhesion of the encapsulation materials.

Peroxides are used in EVA as cross-linking agents. The substances are dosed to ensure that the desired level of cross-linking of the encapsulant is reached during lamination. Since the peroxides are commonly not completely consumed during the cross-linking process, due to intentional over-dosage, too short or too cold lamination, or simply since some of the substance is left over due to normal inhomogeneity, it can also lead to degradation effects of the encapsulant itself or specifically metallic components.

In general, such degradation effects caused by chemical loads in modules can be avoided if interactions of materials are studied in advance, and appropriate procedures are ensured in production to enhance PV reliability.

4.2.2 Electrical loads

PV modules operate on different **voltage levels**, typically up to 1,000 V DC. Further are inside the module different life parts on different voltage levels. The voltage differences within one module are typically <100 V, but they are not negligible. The related electrical fields can have influences on materials, for example, influencing the transport of ions or directly cause degradation.

The **live parts** within one module carry the power, which is generated by all the modules that are connected in one module string. The related current can influence some of the connectors or soldering points especially if they are weak or if they are already damaged due to other influences, for example, weakened materials of the connector wires due to unoptimized processing. In such cases, the electrical loads can cause further damage of the related parts even leading to the total loss of function of the module.

The most common issue of PV modules related to electrical loads is the so-called **partial shading** operational condition. In which case a module is partially shaded and the generated power of the unshaded part of the module or even of the full string is transferred into heat in the shaded area of the module. This can cause so-called hot spots, which can have really severe impacts on the module safety and operation. If the power is not distributed over a larger area but focused, for example, to a small "bad" area or even only a little sensitive spot of one cell, extreme temperatures can be reached. In lab conditions, temperatures of such hot spots with >250 °C have been reported. The caused damages are severe since the polymeric materials cannot withstand these temperatures and also strong mechanical tensions are caused by such temperatures in the module, which can even cause a breakage of the glazing. To ensure safety, this specific situation is addressed as a special test in the type approval and safety standards described in Chapter 8.

There are no specific typical failures reported, which are related to internal electrical loads – besides the hot spot. To the knowledge of the author, there are no degradation effects related to electrical loads that have been explained with classical degradation behavior, by means of a gradual reduction of one parameter over time.

4.3 Classification and mapping of loads

The wish to group load conditions for PV modules for areas or locations with similar conditions grew with growing PV market and increased diversity of module designs and material combinations. Material and module manufacturers would like to use such classes to reduce effort for testing, and qualification of materials and module types and investors as well as insurances would like to use risk maps and load classes for the estimation of expectable degradation rates and service lifetimes.

There are established climate classification systems with long history, such as the most known **Köppen-Geiger system** which has been developed for biology according to the needs of plants and can therefore not be easily used for PV applications. A lot of research is ongoing at the moment to develop useful classification systems and **risk maps** for PV systems, mainly addressing PV modules trying to find measures to identify climates with comparable loads for PV materials. It is expected that such classification systems include the major loads and so also the major degradation processes and effects for PV modules. All the work is done in the area of conflict between two opposing targets: on the one hand – as mentioned in this book several times – each module design with the respective material combination has a very specific degradation behavior and especially specific activation energies describing the relation of accelerated (reliability) testing and normal operation; on the other hand, there are common module designs with expected similar behavior and the product development cycles often do not allow to perform a complete material screening before each change in bill of materials (BOM).

Classification schemes try to first analyze the climatic loads and translate them into microclimatic loads and, second, find areas with similar load conditions. This can, for example, be done separately for each load factor/stressor, which is relatively easy if ambient loads can be directly linked to module loads, for example, for snow loads, as presented in Fig. 4.17. For other loads, more effort is necessary to translate the ambient conditions to microclimatic loads, as it is the case for module temperatures, which can be calculated from ambient wind, irradiation, and location (Fig. 4.18). But to do so, module-specific parameters are required.

Other classification and mapping schemes focus on specific degradation effects or on load combinations. There are also much elaborated classification schemes, including degradation models, as described in Chapter 9, and combine the load data and module degradation behavior to create maps showing expected degradation rates for PV modules (Fig. 4.19) or expected lifetimes or lifetime yields for modules or even systems.

It is important to mention that due to the relevance of the microclimate load data and the specific degradation processes depending on the module design for all the estimations, the classification of one location can be different, depending on the module type and the type of installation.

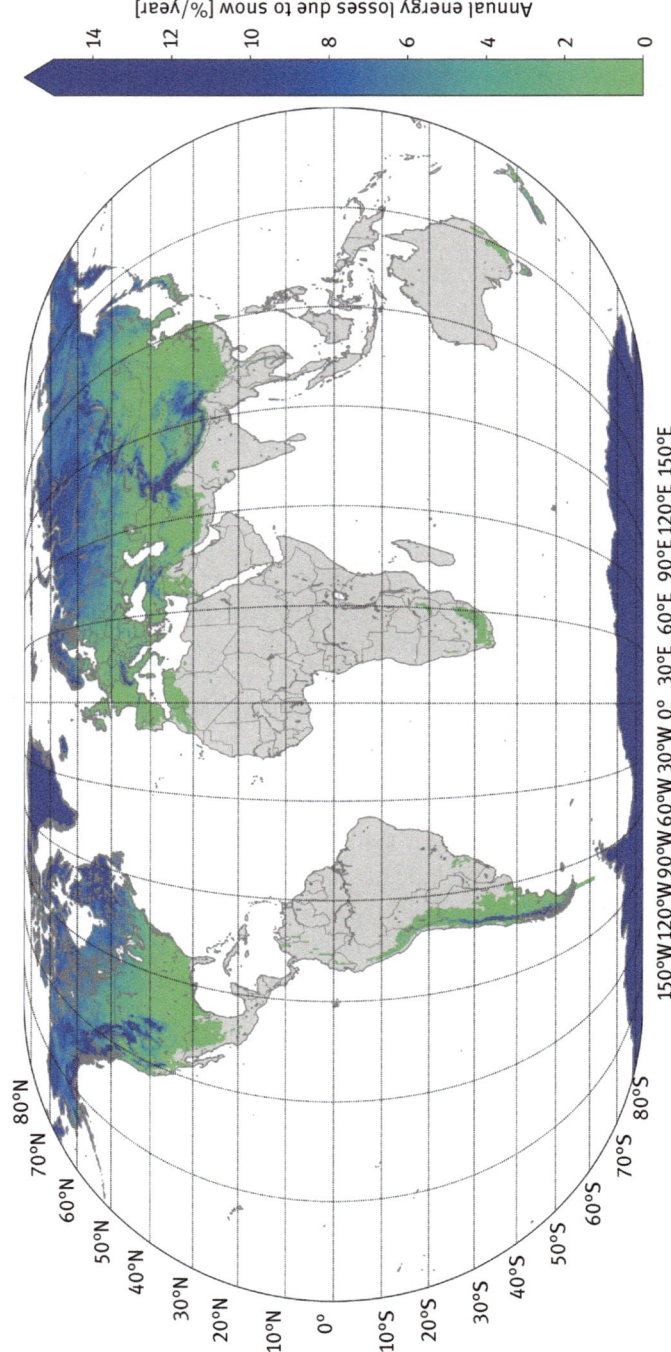

Fig. 4.17: Global map showing energy losses due to snow loads for PV modules and related yield losses. Source: Julián Ascencio-Vásquez, LPVO-UL.

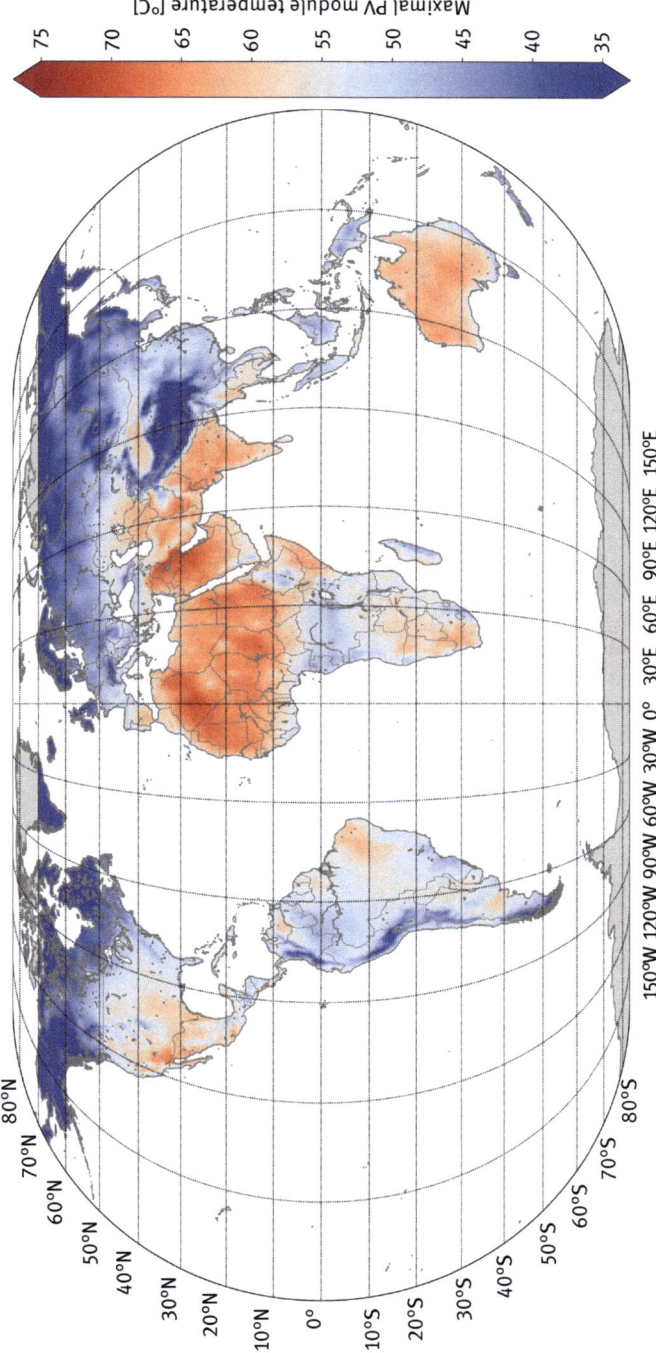

Fig. 4.18: Global map showing the maximum expected module temperatures. Source: Julián Ascencio-Vásquez, LPVO-UL.

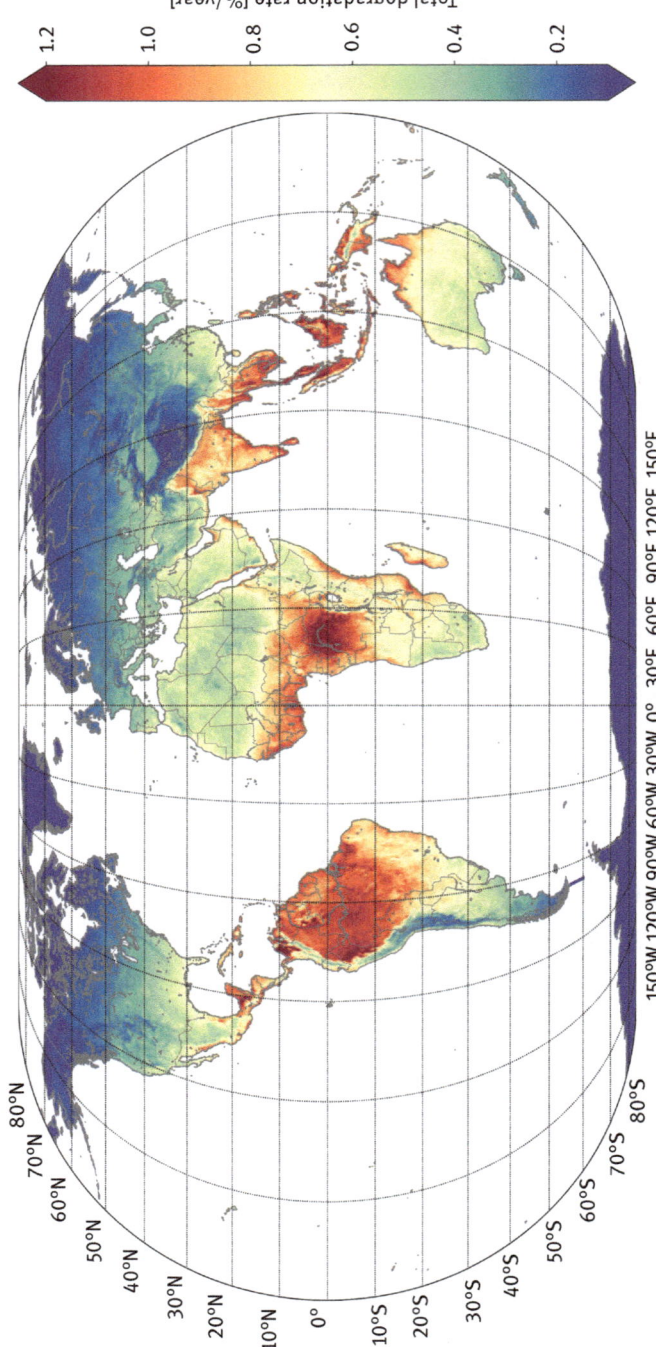

Fig. 4.19: Global map showing expected module degradation rates calculated with degradation models and including load maps. Source: Julián Ascencio-Vásquez, LPVO-UL.

Karl-Anders Weiß
5 Accelerated aging tests

All environmental simulation tests, which allow an accelerated analysis of material degradation processes through intensified environmental stress levels compared to real life, can be summarized under the notion of accelerated aging tests.

These types of tests are especially useful for products with long service lifetimes, such as photovoltaic modules. As warranties for PV modules cover up to 30 years of service life, accelerated aging tests are of high economic importance for PV module manufacturers. As PV manufacturers aim for utilizing materials, which are both economically and technically suitable, accelerated aging tests on full PV modules as well as on separate materials and material combinations are necessary. It is in the interest of the customer as well as the investor and the insurance that the product achieves the expected service life without major performance losses. The aim of accelerated aging tests is to achieve reliable results within a very short period of time and with minimum costs to keep pace with the rapid innovation cycles. In order to ensure the transferability of the testing result on real-life performance, a specific adaptation of the test conditions to the specific materials and stress factors is indispensable. Otherwise, climatic conditions may be induced, which either exceed or undermine those under standard operation, and this may subsequently lead to altered physical or chemical degradation processes. Thus, the processes that occur under standard operation should be accelerated, but at the same time, simulated as close to reality as possible.

The acceleration factor a is defined as

$$a = \frac{t_{test}}{t_0} \tag{5.1}$$

with the testing time during accelerated aging (t_{test}) and the exposition time under standard operation (t_0), both of which result in an identical (material) degradation. The dependency of the acceleration factor on the testing conditions and the degradation factors is described by time transformation functions, which can be used for the estimation of the service life of a product.

In general, there are three ways to achieve the required acceleration:
1) constant high stress level of the degradation factor;
2) increasing & variation in the intensity of the stress factor; and
3) increasing the temperature to accelerate the degradation process.

In all cases, the potential of inducing unrealistic processes and subsequently unrealistic material degradation processes exists. By 1) relaxation phases, which may occur in reality, for example, during the night or times needed for transportation processes, can be neglected. Option 2) has the risk of altering the dose–response relationship significantly, for example, if the limiting reaction process has no enough

https://doi.org/10.1515/9783110685558-005

time or reaction partners to react. Possibility 3) bears the danger that the different processes have different temperature dependencies, and thus an alteration of the reaction equilibrium may occur. Additionally, temperatures may be too high, and this way, nonrealistic conditions may be achieved. Therefore, the specific conditions should be evaluated thoroughly before choosing the most suitable conditions.

In the following, the basic equipment used to perform accelerated aging tests for PV modules and materials are presented as well as the processes to develop appropriate tests. The described approach is more or less application neutral and can be easily transferred to other applications since the relevant loads and the related testing equipment is comparable for all products used in outdoor applications.

5.1 Light sources

As described in Section 4.1.1, the degradation process of PV modules depends highly on the irradiation level during the aging process. Thus, the selection of a light source and the related spectrum is of highest relevance for accelerated aging tests. In addition, all irradiation comes along with the transfer of energy. Tests with irradiation usually also impact the temperature of the sample and this has to be included in the planning of tests. Different light sources have different ratios of the shares of the different parts of the spectrum, as shown, for example, in Fig. 5.1. The generally used

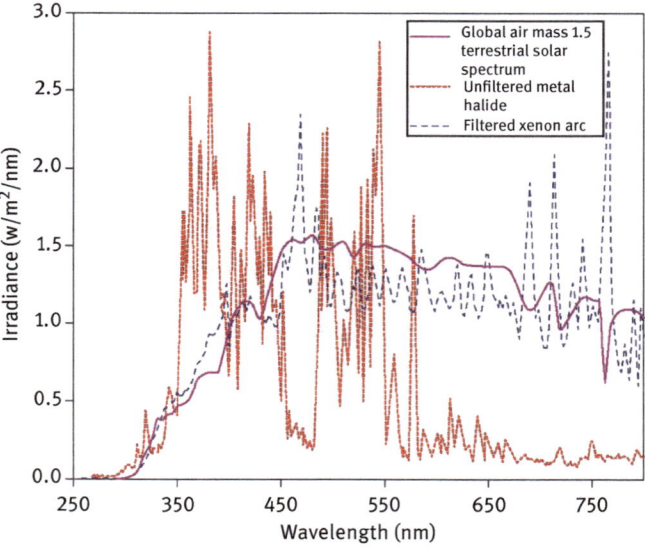

Fig. 5.1: Spectral distribution of the irradiance of the sun (AM 1.5 spectrum, magenta), a metal halide lamp (red), and a xenon lamp with a filter (blue) [47].

spectral "sections" are usually ultraviolet (UV), visible light, and infrared (IR), but a more detailed analysis is necessary to compare and rate light sources.

Ideally, an accelerated aging test would use the full solar spectrum, for example, an AM 1.5 spectrum as shown in Fig. 5.1, in order to eliminate issues related to altered spectral distributions during the aging process. But due to the occurrence of overheating of the sample, especially in case of high irradiance levels, as well as the predominant impact of short wavelength irradiation on the degradation processes, light sources are usually chosen according to their UV levels.

The efficiency with which an incoming photon may induce damage in a material varies between different materials depending on color, surface conditions, and other factors, but is exponentially related to the shortening of the wavelength of the incident photon.

The energies and corresponding wavelengths that are required to induce bond breakages of the most important organic groups are given in Fig. 5.2.

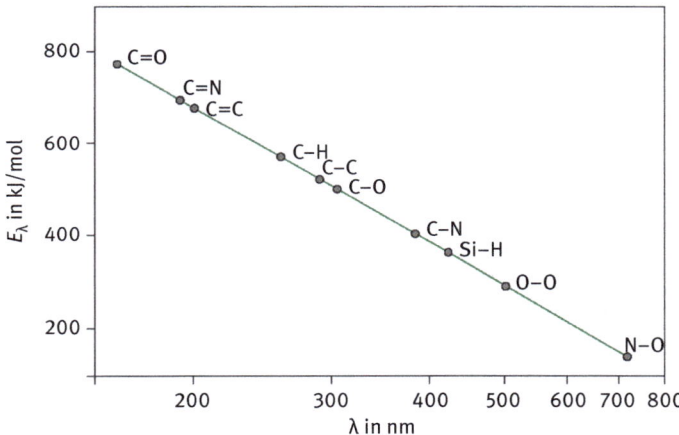

Fig. 5.2: The energies and corresponding wavelengths that are required to induce bond breakages of the most important organic groups, according to Lechner et al. [48].

Different types of light sources can be used for accelerated aging tests. These light sources fulfill the described requirements to different extents. In case the spectral sensitivity of the sample is unknown, it is suggested to use a light source with a spectral distribution close to the solar spectrum. Xenon light sources with filters are especially well suited for this application, as shown in Fig. 5.4.

For the analysis of polymers, it is important to achieve very similar sample temperatures between different samples while at the same time, the UV range of the spectrum is aimed to be enhanced. For this objective, light sources with a relatively high UV portion such as metal halide lamps or fluorescence tubes are most suitable (Fig. 5.3). For choosing the most suitable light source for a UV test, it is very important

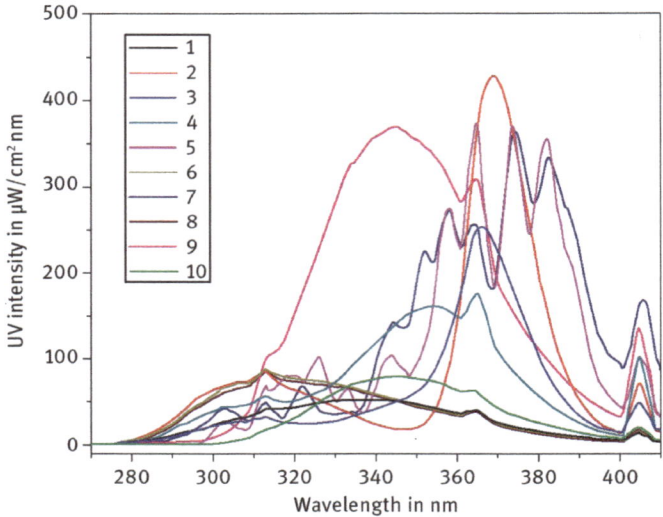

Fig. 5.3: Comparison of the intensity and the spectral distribution of 10 different UV light sources used in a round robin between different test laboratories [49]. The light sources are UVA fluorescence tubes (1, 9, 10), UVA + UVB fluorescence tubes (2, 3, 4, 6, 8), and metal halide lamps (5, 7).

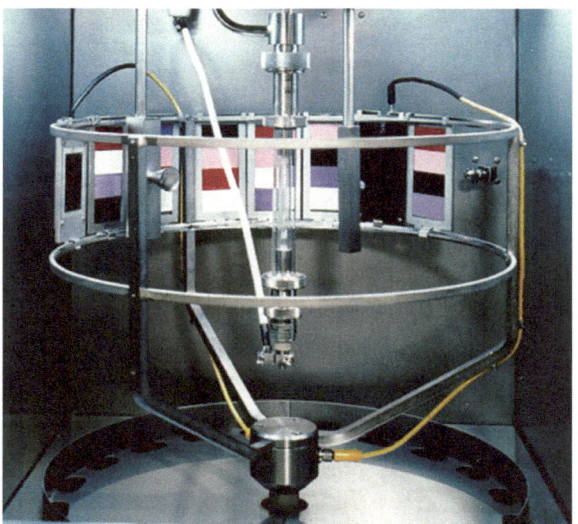

Fig. 5.4: Xenon light source with filter mounted in climatic cabinet with turning sample holder together with some samples and white standard temperature (WST) and black standard temperature (BST) sensors. Source: Atlas Material Testing Technology GmbH.

to know the spectral distribution as the applicable norms such as ISO 4892, IEC 61215, 10.10 allow a wide range of possibilities as shown in Fig. 5.3.

There are a lot of different types of equipment available for materials testing. The smaller the samples are, the bigger is the variety.

For the analysis of large samples, the spectral and spatial homogeneities of the light source are of great importance in order to allow for a comparability of the results. The temporal and spatial homogeneities of the spectrum of solar simulators are, for instance, classified in the standard IEC 60904-9. The light from a solar simulator is controlled in three dimensions:
1) spectral content,
2) spatial uniformity, and
3) temporal stability.

A solar simulator meeting class A specifications in all three dimensions is referred to as a class A solar simulator, or sometimes a class AAA (referring to each of the dimensions in the order listed earlier). IEC 60904-9 classifies the classes of solar simulators according to the description in Tab. 5.1.

Tab. 5.1: Specification of the different dimensions for solar simulators according to IEC 60904-9.

Classification	Spectral match (each interval)	Irradiance spatial nonuniformity	Temporal instability
Class A	0.75–1.25	2%	2%
Class B	0.6–1.4	5%	5%
Class C	0.4–2.0	10%	10%

Solar simulators of very high precision, especially in temporal stability, are usually not necessary for accelerated aging test. To provoke degradation effects, exposure usually takes several hours up to several thousand hours, so instabilities in the subsecond level are not relevant. For the aging tests, it is important that the relevant part of the spectrum is included in the radiation of the light source and that the irradiation dose is monitored well. For analytical purposes, like the determination of IV curves as described in Section 3.2.1, all three dimensions have to be carefully controlled.

Especially for tests with large samples like full-size modules, it is not easy to ensure spectral and spatial uniformity. These issues are usually easier to implement with fluorescence tubes (Fig. 5.5) or other sources delivering uniform radiation of and to an area, in comparison to point light sources.

There are solar simulators available, which reach class AAA on very big areas, but, as mentioned earlier, this is not necessary for degradation testing. When such equipment is needed, it is important to ensure that the conditions are reasonable. Typical

Fig. 5.5: UV exposure of full-size PV modules using fluorescent tubes.

powerful light sources produce a lot of heat and the sample has to be protected from this influence (Fig. 5.6). One possibility for that is the so-called cold sky, which uses an air flow perpendicular to the light beam to remove the heat reaching the sample.

Fig. 5.6: Steady-state solar simulator reaching class BBA for illumination of big or several samples using metal-halide light sources and a "cold sky" to reduce IR irradiation. Source: PSE AG.

5.2 Climatic cabinets

Climatic cabinets are all devices which are able to produce and maintain specific environments for accelerated aging tests of test samples. The most simple form of a climate chamber is a device, which regulates the temperature level, it is also referred to as an "oven." Two examples of these "ovens" which are suitable for very different temperature ranges are shown in Fig. 6.1. More sophisticated devices allow for an additional humidity regulation. Especially, if tests shall address effects which are not clearly and exclusively linked to temperature loads, controlled climatic conditions are crucial. Then it is to be ensured that the relevant and desired processes will take place. This requires cabinets, which can also control other loads, for example humidity. The climatic cabinets are typically characterized according to the possible tests they perform. These parameters include, for example, temperature range, humidification power, how fast they can change the temperature in loaded operation, and the size. An example of such a smaller climatic cabinet which is typically used for material testing is shown in Fig. 5.7, and an example of a bigger cabinet adapted to PV module testing is shown in Fig. 5.8.

Fig. 5.7: Climatic cabinet, which allows the regulation of temperature and humidity inside the test chamber. Source: Weiss Umwelttechnik GmbH.

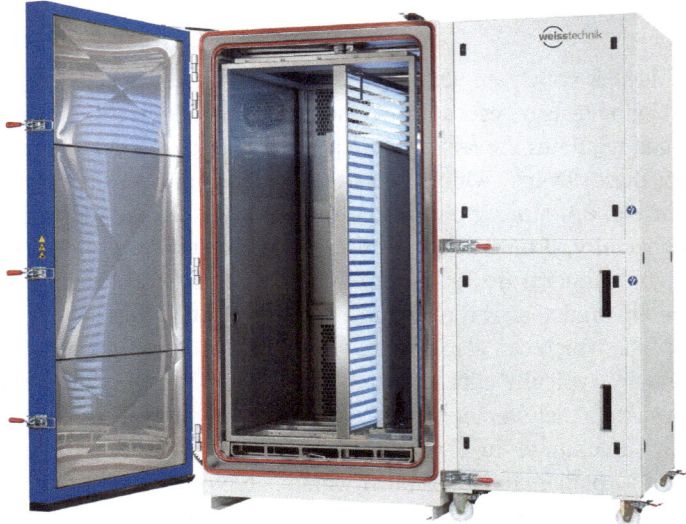

Fig. 5.8: Climatic cabinet adapted to PV module testing. Source: Weiss Umwelttechnik GmbH.

For many tests, it is important to ensure that all relevant loads are applied to the sample in parallel or at least it is ensured that all reaction partners are available at one time. Especially if polymer degradation processes are included, the availability of UV radiation and humidity in parallel is necessary. A climatic cabinet, which offers radiation loads and controls humidity and temperature, can also be referred to as weathering testing instrument. An example is given in Fig. 5.9.

Within this climatic chamber type, the exact regulation of the climatic conditions is rather difficult, as the light sources cause an intense heating of the chamber, which in turn needs to be controlled by cooling measures. Generally, the homogeneity of the chamber conditions throughout the whole chamber is a critical point, especially in case of rapid temperature cycling and at high temperatures and high humidity with high irradiation.

As described in Section 5.1, homogeneity with regard to time and area is often a critical point during accelerated aging tests. Additionally, it is important to differentiate between the ambient atmosphere (macroclimate), which is the controlled environment in the test chamber and the microclimate, present within or directly at the sample. As the parameters of the environment are amplified compared to real-life conditions, the resulting microclimate is also often significantly altered, which may lead to a significant acceleration of some of the processes. While this is a generally desired phenomenon in order to achieve short testing times, it is of great importance to know the exact level of amplification. Thus, precise measurements of the different parameters of the microclimate are absolutely crucial for successful and reliable accelerated aging tests. The determination of the sample temperature is thereby

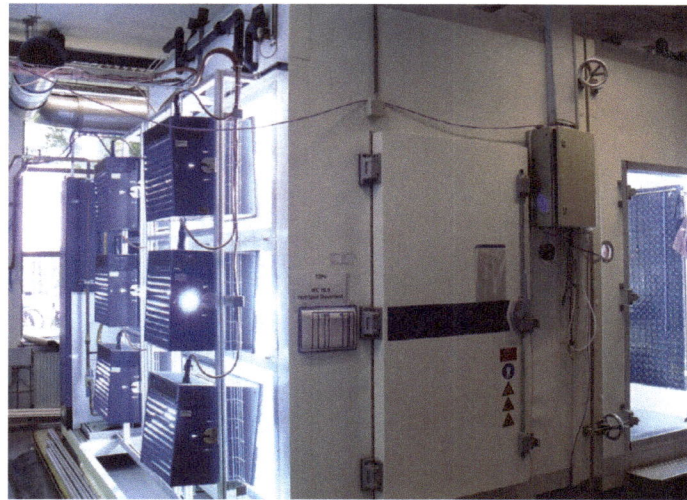

Fig. 5.9: Double climatic chamber with solar simulator (class B according to IEC 60904-9).

most difficult during an accelerated aging test, which includes irradiation of the sample. To address this problem, the so-called black standard temperature (BST) or white standard temperature (WST) sensors are available. The optical properties and the related surface temperatures of samples usually lie in between the two extremes of the BST and the WST. The direct measurement of the sample temperature can also be difficult, as the measurement itself can impact on the conditions at the sample [50]. The microclimate should be considered especially in the case of several samples in one test with the aim of comparing the results.

Corrosive environments have a major impact on the degradation of PV modules, as described in detail in Section 4.1.5. Therefore, accelerated aging tests, which simulate the degradation characteristics in marine or other highly corrosive environments, are of great importance, even though they are relatively new compared to established tests like UV or damp heat aging. As salt mist is one of the most important factors for this type of degradation, salt mist chambers as shown in Fig. 5.10 can be used for accelerated aging. Not all salt mist or salt spray cabinets are able to perform all the possible tests, which reduce the flexibility in test design significantly. Generally speaking, it is not really common to adapt corrosion tests to the specific application or load situation as it is established for other loads, such as T, UV, or humidity. Therefore, corrosion test chambers are often very much optimized for specific tests. More flexible chambers are not common and require a lot of expertise to operate; this is, for example, due to the requirement to ensure a uniform distribution of the humidity and salt in the cabinet, especially when samples are installed. When new or adapted tests are developed and applied, usually a lot of adjustments are necessary. Typically, the deposition of the salty fog and the corrosivity is monitored to analyze the operation of the device.

Fig. 5.10: Salt mist chamber at Fraunhofer ISE.

There are a lot of different standardized tests which can be applied; an overview is given in Chapter 8. These standardized tests do not directly correlate with loads found in outdoor operation and so results are to be seen as sensitivity indicators helpful for relative comparisons.

Other corrosivity tests can be performed using specific corrosive gas chambers, such as ammonium test chambers.

5.3 Mechanical and hail testing equipment

Mechanical testing of materials is typically performed with equipment similar to the standard testing machines described in Section 3.1.6 or with dynamic mechanical analysis tools as described in Section 3.1.3.

Full-size modules and laminates alike need to be tested to ensure the stability of the connection of the components and to analyze their reaction to external mechanical loads caused by, for example, wind or snow loads. The goal is to ensure a uniform load on the sample surface in these tests, which can be ensured by adapted load bags using air or water pressure. Another way to nearly reach uniform loads is using a reasonably high number of large, mechanic vacuum stamps to apply the load to the module. The latter is the most common setup of such tests. An example of such a setup is shown in Fig. 5.11.

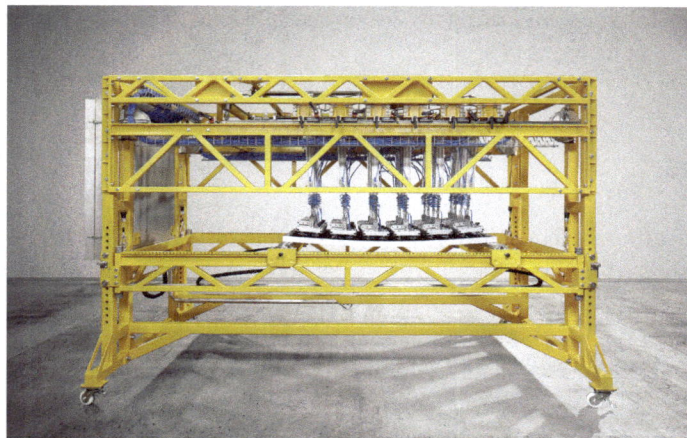

Fig. 5.11: Mechanical load testing of a PV module. Source: PSE AG.

Hail testing, as a very special version of mechanical impact testing, is commonly performed with the so-called hail guns (Fig. 5.12) using ice balls or polymeric balls with different diameters as bullets. There is also a variety of impact velocities defined by different standards. The ball size and weight define together with the velocity the hail class they are representing.

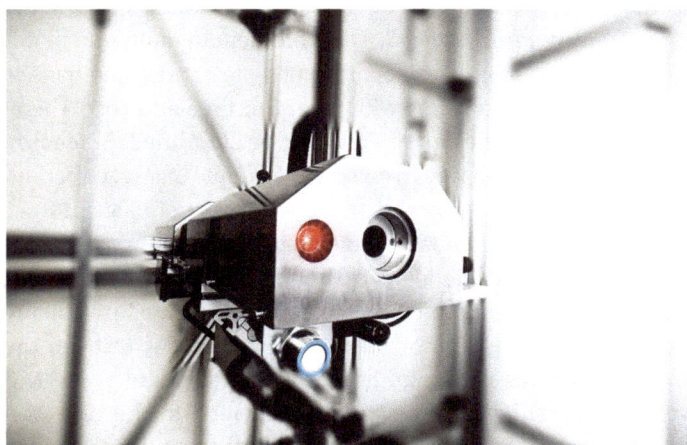

Fig. 5.12: Hail gun setup for testing of PV modules. The laser supports the accurate positioning. Source: PSE AG.

In addition to hail tests, there are also so-called module breakage tests to investigate the effect of massive impacts on a module to ensure safety. These impacts cannot be seen as degradation effects, and they are therefore not described here.

Specific approaches for abrasive testing are described in Section 6.2.1 since they are mainly relevant for glazing materials.

5.4 Procedures

The amplification of the different loads during an accelerated aging test results in various risks and difficulties, which were described earlier in this chapter. These risks have to be taken into account for the design of an accelerated aging test procedure or sequence. Therefore, a comprehensive experimental design is required. This experimental design has to consider all available information with regard to the sample, such as

– specific temperatures which should not be exceeded, or
– temperatures at which physical transitions (T_G, T_M) occur, or
– known interactions with other materials,

which will be present in the final application.

On this basis, with the help of a statistical experimental design, an experimental plan can ideally be developed aiming to evaluate the dose–response relationships between the different parameters and degradation indicators. Target is often to limit effort and have as few as possible separate aging tests. Therefore, complete parametric test plans, according to them, every single parameter has to be varied keeping the other parameters constant, are very rarely applied. To fulfill them, a multitude of different climatic conditions and a large number of separate aging tests arise and have to be performed. This may easily lead to problems with regard to time and cost. Therefore, this ideal statistical experimental design often has to be simplified. Screening tests for a sensitivity analysis are a feasible way to do so. An example of a set of parameters for an experimental test sequence for polymeric materials is given in Fig. 5.13.

These controlled artificial climatic condition tests aim at identifying the sensitivity of the samples with regard to specific loads or combinations of loads as well as at determining thresholds, which should not be exceeded. This procedure usually starts off with climatic conditions of well-established accelerated aging tests for the type of sample investigated – such as the damp heat test for PV modules at 85 °C and 85% r.h. – followed by a variation of the single parameters. In the case of a planned utilization of the aging test results for modeling, it is of great importance to obtain enough data points in order to achieve a reliable base for the calculation of the kinetics and the dose–response relationship. To do so, it is important to include different intensity levels of all relevant loads to be able to determine acceleration factors, which are prerequisites for translating accelerated indoor lab tests to outdoor operation.

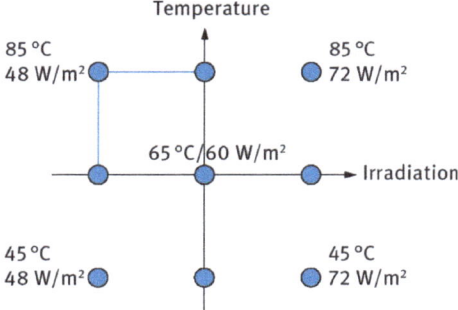

Fig. 5.13: Exemplary experimental design with a variation of the parameters temperature and irradiation. Goal of the experiment was the investigation of sensitivities of different polymeric materials towards the different stressors and the determination of acceleration factors.

The selection and the adjustment of specific parameters for accelerated aging tests with regard to expected environmental conditions during the service life in application is called test tailoring.

Karl-Anders Weiß, Elisabeth Klimm

6 Reliability testing of materials

In this chapter, some of the most important equipment items as well as approaches and procedures for the reliability testing of materials for photovoltaic (PV) applications are presented. It must be mentioned that for reliability testing, it is important to know the real occurring loads for the tested materials as well as the relevant operational conditions. This is also a major difference to accelerated aging testing, which is in the first approach independent of the operational conditions even when the equipment is similar. Therefore, the specific tests in reliability testing can be very similar as in the accelerated aging tests, but the definition and interpretation are more difficult. They require more information, and thus the results are also much more meaningful for the application. Furthermore, a general procedure of qualification testing of materials is given as well as short descriptions of specific requirements and conditions of different components. There are also standards for the qualification of materials for PV modules available (see Chapter 8). Nevertheless, these are by nature not addressing the complete issue of reliability.

6.1 Equipment

In general, all pieces of equipment described in Chapter 5 can be used for material testing. For some tests, it is simply easier and cheaper to do materials testing with equipment of smaller size which is fine for materials but not reasonable for modules.

Chapter 4 described the general importance of temperature for the durability of materials. A very basic and simple stability testing can be performed in ovens. Figure 6.1 shows the ovens for material aging, which are suitable for accelerated aging tests at different temperature levels.

In addition, climatic chambers that allow temperature, humidity, and UV testing – as described in Chapter 5 – are used for reliability testing of materials for PV applications. Since for reliability testing, combined loads often have to be simulated, for example, combined exposure to UV loads, temperature loads and humidity loads, more flexible, sophisticated, and multifunctional equipment, such as climatic cabinets, with the possibility to additionally apply UV loads, are often necessary.

As further described in Chapter 4, the water vapor diffusion and oxygen permeation through polymeric materials in PV modules are of great importance for the durability of the whole module. In order to test the specific permeation properties

https://doi.org/10.1515/9783110685558-006

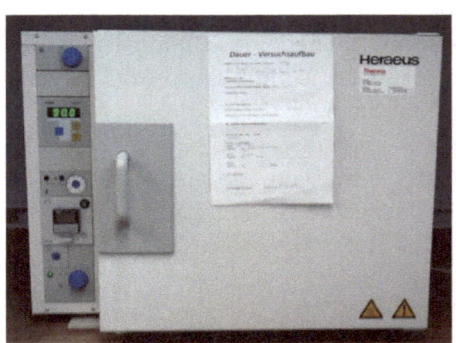

Fig. 6.1: High temperature testing cabinets for material aging which allow accelerated aging tests up to 200 °C (left) and 800 °C (right).

of backsheet and encapsulation material, special test chambers were developed, which allow a time- and temperature-dependent measurement of the permeation rate. In Fig. 6.2 such a chamber, which utilizes a mass spectrometer as a detector, is shown. This is not a reliability test by itself. The setup rather delivers relevant data, for example, of the backsheet and the encapsulation material, for the definition of test parameters. It is of special importance for materials in the module which are separated from the ambience by the barrier material like the encapsulant or the backsheet.

Fig. 6.2: A permeation measurement climatic chamber which utilizes a mass spectrometer as detector.

6.2 Procedures

Figure 6.3 shows the general procedure of reliability testing for new materials in PV applications. Exactly, this procedure is often used for the qualification of materials during the development process. It involves an initial stage, in which the new material is characterized. These methods can comprise transmittance measurements, tensile testing, or Raman spectroscopy, depending on the material/component to be investigated. New materials are subject to different accelerated aging tests or test sequences, such as damp heat (DH) testing, UV testing, or combined UV and DH aging. The subsequent characterization routine is identical to the one before the aging procedure, in order to compare the material characteristics before and after aging. If materials show an acceptable aging behavior at this initial stage, small PV modules are manufactured.

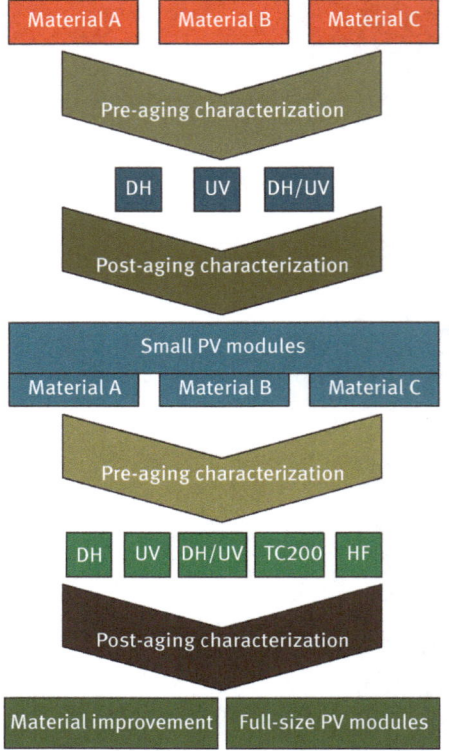

Fig. 6.3: General procedure of material testing for PV application.

These mini-PV modules include the new materials in addition to all other PV components as in a standard manner. That way testing for the specific material interactions, which are to be expected in a full-size PV module, is possible. Especially the material interactions between the solar cell metallization, the encapsulation, and the backsheet material are of great interest for investigation, since corrosive degradation byproducts may dramatically change the degradation pathways and thus the speed of the degradation. Therefore, it is important to ensure that realistic microclimatic conditions occur in the weathering tests for the material of interest, as well as oversimplification of the test laminate design has to be prevented. The mini-PV modules (or test laminates) are subject to more comprehensive accelerated aging tests such as humidity freeze and thermal cycling. They are usually not included in the initial phase, as these tests are more adequate when boundary layers, as in laminates are present, are stressed by mechanical load. These tests can also be adapted to the expected climatic and operational conditions for the use of the materials; for example, materials for roof integrated modules in southerly countries should be tested at higher temperatures than materials for the application in open rack installations in Germany.

For some materials, reliability tests can also be performed on single material samples; hence, minimodules are not necessary if the functionality is not influenced by the interaction with other components like the coatings of the glazing material. Also, the materials in small modules undergo a pre- and post-aging characterization routine, which may include analytical tests as described in Section 3.2. Depending on the outcome of these tests and characterization measurements, the second phase will either be followed by material improvement activities or, in case of well-performing minimodules, full-size PV modules can be manufactured and tested.

6.2.1 Glazing materials

Elisabeth Klimm

The glazing of PV modules has two essential functions to fulfill. On the one hand, the glazing should offer a protective function against environmental influences. This includes protection against moisture, hail, abrasion, and against surface-attacking substances. It should also serve as electrical insulation. On the other hand, as much light as possible should be transmitted in order to achieve a high energy yield. Most crystalline silicon modules employ glass cover plates for the provision of structural strength and the encapsulation of the cells. Hence, glass is the single largest component by mass in most solar modules in production. Low-iron solar glass is favored for this application due to a range of beneficial properties, including its strength and rigidity, excellent transmittance of photon energies, multifaceted coatability, environmental stability (i.e., weathering and UV resistance), and relatively low costs. Solar glass is typically tempered to withstand mechanical stress as well as inhomogeneous temperature distributions. White soda–lime–silica glasses are commonly used, as well as occasional white borosilicate glasses. They further typically exhibit a low iron content (as low as 100 ppm compared to 1,000 ppm in regular soda lime), which improves light transmission and thus increases efficiencies of the PV module. There are also polymer-based glazing materials, for example using polycarbonate (PC) or polymethymethacrylate (PMMA), but only with a negligible market share at the moment. For polymeric materials UV testing is much more relevant.

Many glazing materials are equipped with functional coatings or structured surfaces to improve specific properties. Most relevant are antireflection (AR) coatings to improve the light gain and antisoiling (AS) coatings to reduce soiling effects or make cleaning easier. Testing of these coatings is included in the section since the coatings can only be analyzed together with the glass substrate. The relevance of the different loads depends on the specific material of the glazing and the coating and the application.

The following analytical tools are typically used (depending on the available equipment and the specific purpose of the test) to analyze glazing materials:

- Transmission measurement, for example, by Fourier Transformation Infrared (FT-IR) spectroscopy to determine the optical properties of the material
- Atomic force microscopy (AFM) to analyze the surface properties
- Contact angle measurements by optical tensiometer to analyze the wettability of the surface, especially if functional coatings have been applied

Due to the specific properties of glazing materials, they can usually be tested independently; since there is no relevant degradation inducing interaction with other components expected. The different functionalities and the specific position of the glazing as outer layer make special tests essential for the reliability assessment. On the one hand, **condensation** testing directly addresses the reliability of the surface, and is therefore of special importance for coated glazing materials. Another relevant test addressing the stability of the surface is **humidity freeze testing**, which can show effects on porous materials. Also high humidity at high temperatures, with **damp heat testing**, can discover surface degradation effects. All tests can invstigate possible glass corrosion. Glass corrosion is caused by the leaching of oxides. A formation of a brittle white layer on the surface makes the glass opaque and can cause small cracks. To avoid this fatal failure it is recommended to perform reliability testing of glazing materials, including all typical PV module tests plus the special tests as described in this chapter.

Degradation is usually determined by analyzing the surface conditions and the transmission of the glass. On the other hand, there are tests addressing abrasive and soiling loads, which are typically only applied to glazing materials. As described in Section 4.1, these loads are very location dependent and therefore the testing has to be adapted to the specific requirements of the planned site. Since the relevant equipment is only necessary for glazing testing, it is also described briefly in this chapter.

Experimental investigation of soiling effects is made with different custom-made setups at different institutions, since there is no standardized approach available, yet. Available "dusting devices" for standardized dust testing as in the IP tests so far aim at bringing large quantities of dust into an (electrical) system without a qualitative distribution pattern. The new **soiling tests** are to be seen as a functionality tests detecting, for example, performance and/or degradation of functional surfaces. Hence, soiling testing itself is not a reliability test. Most soiling tests are done with dry dust. The goal is to quantify the soiling losses related to a reproducible and defined soiling load. In comparison to abrasion testing is homogeneous dust deposition onto surfces, such as anti-soiling (AS) coatings on solar glass simulated. Soiled samples are subsequently analyzed with optical and gravimentrical tools.

There are approaches to combine soiling testing directly with artificial weathering, such as light, temperature, or humidity exposure. If such artificial weathering is used to create soiling patterns and the exposure is for sure not degrading the samples surfaces a fast, reliable, and reproducible soiling method can be defined. The assessment of soiling effects is done by transmittance measured via FT-IR spectroscopy before and after artificially soiling the sample surfaces. In parallel are the deposited amounts of dust weighted. Plotting the deposition in gram per square meter of the transmittance loss, a dust-specific graph with usually linear relationship is found.

Degradation effects of surface properties can be discovered with artificial soiling tests. Figure 6.4 shows the application of a soiling test as characterization before and after accelerated aging.

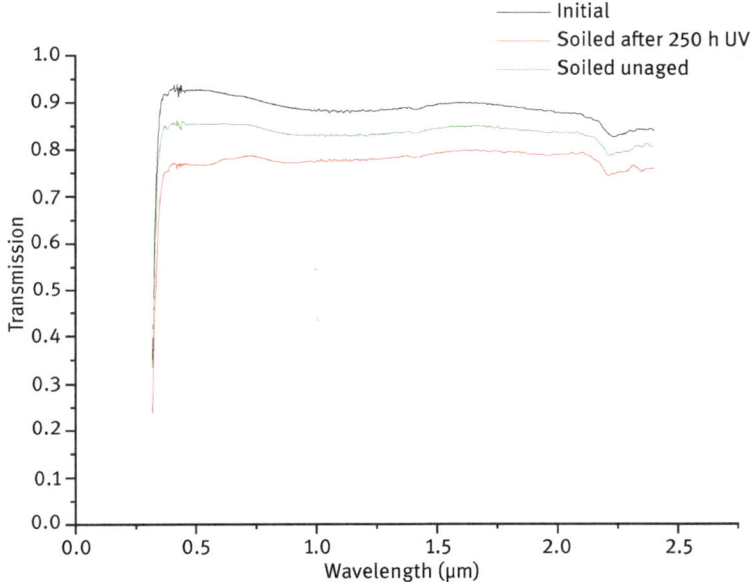

Fig. 6.4: Transmission of one sample of coated glass in the initial (clean) state and after artificial soiling and soiled after exposure to 250 h of UV testing.

As mentioned earlier, soiling testing is not related to irreversible degradation of surfaces, while during abrasion particles impact the surface irreversibly. Or particles alter the surface through chemical bonding irreversibly.

Abrasion testing can be done using different standardized tests such as Taber abraser, sand blowing tests or sand trickling tests. If such tests are applied, it is recommended that sand is used with similar properties as the soil in the location where the material shall be used to ensure relevant results. Testing with different materials can cause different degradation, as shown in Fig. 6.5, with sands from two test sites. In this test also different drop heights of the trickling sand were used to simulate the different typical wind speeds of the two sites.

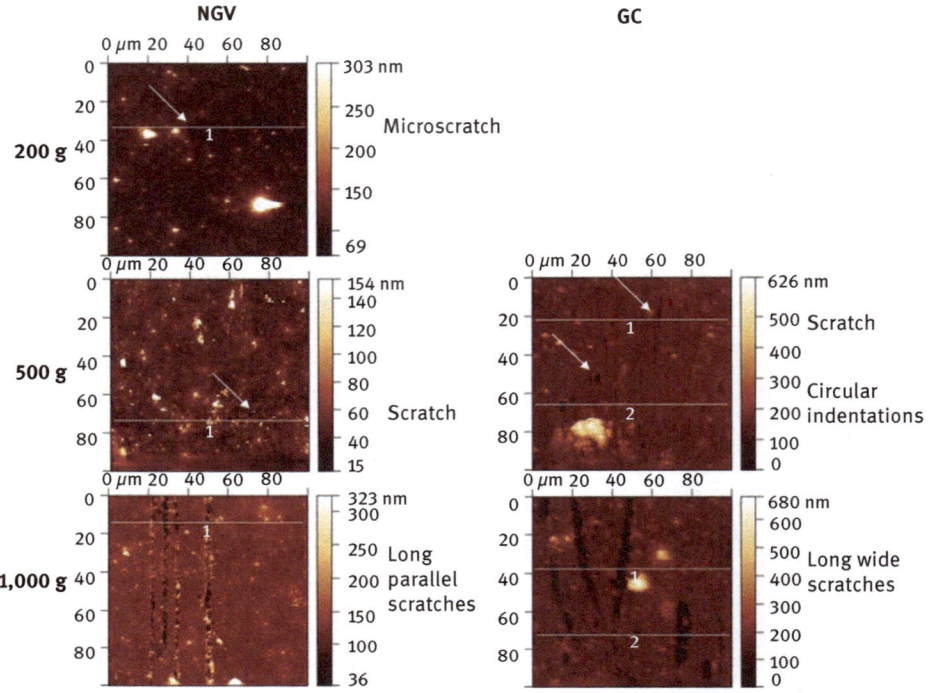

Fig. 6.5: Surface effects of coated glass samples measured with AFM after sand trickling test with different amounts of sand and different sand types from test sites in the Negev desert, Israel, and on Gran Canaria, Spain.

6.2.2 Encapsulation materials

The encapsulation polymers have to fulfill multiple functions in the PV module, and all these properties have to be considered when the reliability is analyzed. The basic requirements are that the encapsulant is highly transparent to let as much light as possible passing to the PV cell (optical coupling). Further, the encapsulant ensures the mechanical connection of the module laminate consisting of glazing, cells, encapsulant, backsheet, and electrical connectors (mechanical coupling) and protects the cells from mechanical impacts. Thus, all relevant parameters such as transparency, adhesion, and elasticity should be included in the testing. Analytical methods have to be selected in a way that they are able to provide proper information on the relevant material property. In the case of encapsulation material, the following tools are typically used, in dependence of the available equipment and the specific purpose of the test:

–   Transmittance measurement, for example, by FT-IR spectroscopy is used to determine the optical properties.

- Raman spectroscopy/Raman microscopy to analyze the bondings of the material, and gain information on the chemical composition and e.g. on the degree of cross-linking [85]. In confocal mode, this method can also deliver information with lateral resolution and also generate depth profiles.
- Fluorescence measurements are mainly used to present insights on degradation products, since double bonding generates the measured fluorescence signal.
- DMA and tensile testing are used to analyze mechanical and elasticity properties
- Gel content analysis is used to analyze the degree of cross-linking.
- Tensile testing and peel testing are used to analyze the adhesion strength of encapsulants to glazing, cell, or backsheet.

Encapsulants are polymeric materials and are typically sensitive to elevated temperature, humidity, and UV radiation. Thus, all the climatic load factors have to be included in the testing. Since the encapsulation material is situated inside the module laminate, separated by backsheet and glazing from the ambience, the microclimatic conditions for the material differ significantly from the ambient conditions. This fact is relevant besides the normal operation but even more in accelerated tests with usually higher intensities. Therefore, there are only few reasonable tests, which can be performed on samples consisting of just the encapsulation material. Encapsulants are typically under test using small laminates of glazing, cell, encapsulant, and backsheet. The size of these laminates should be chosen in a way that edge effects do not dominate the conditions in the sample. It is recommended to choose a size of at least 20×20 cm^2. When defining the testing conditions for the samples, it is important to be aware of the effect of the different load intensities (T, UV, rh) on each other. High irradiation levels cause high sample temperatures and induce low relative humidity values. For this reason, it is essential to estimate the microclimatic conditions before starting the tests. Possibilities to calculate or roughly estimate the microclimate for these types of samples are similar for full modules and are described in Section 7.2 in more detail.

It is recommended to test the sensitivity to each load factor/stressor and beyond to perform combined testing, including several load factors in parallel or series. An approach could be, for example, exposing samples to UV under controlled temperature and humidity conditions to ensure that the sample does not reach excessive temperatures with too low humidities in order to cause the targeted chemical degradation. Further, an exposure to temperature cycles could be applied to test for changes of the mechanical properties.

Table 6.1 shows the results of an extensive screening test of combinations of four encapsulation materials with five different backsheets. Minimodules have been built for the testing and were exposed to 400 h of UV testing at 65 °C, 1,000 h of DH testing, or combined UV–DH testing for 1,000 h, respectively, and were analyzed and rated afterward.

Tab. 6.1: Results of degradation screening tests for different combinations of encapsulation materials: EVA fast cure (EVA fc), PVB, thermoplastic silicone elastomer (TPSE), and an ionomer with different backsheets: polyvinyl fluoride (PVF)–polyethylene terephthalate (PET)–PVF (TPT), polyamide (PA, AAA), PVF-PET-PA (FPA), PET, and polyvinylidene fluoride (PVDF)-PET-PVDF. The rating includes results from UV, damp heat (DH), and combined UV–DH testing. In each test, the samples have been rated for "no effects" (5 points) to "very strong effects" (1 point). For the final rating, the results of the different tests have been summarized.

	EVA fc	PVB	TPSE	Ionomer
TPT	11 4 + 4 + 3	12 5 + 3 + 4	10 4 + 4 + 2	14 5 + 5 + 4
AAA	13 5 + 3 + 5	7 4 + 2 + 1	8 4 + 3 + 1	13 5 + 4 + 4
FPA	9 3 + 4 + 2	–	–	–
PET	9 3 + 4 + 2	–	–	–
PVDF/PET/ PVDF	12 4 + 4 + 4	–	–	–

6.2.3 PV cells

PV cells are typically qualified according to their electrical properties to generate electric energy from incoming solar energy. There are no reliability tests specifically for PV cells in the context of established module technology. Some effects can occur, such as degradation of the cells' AR cell coating or of the metallization as well as cracking and breakage of the cells. All these effects only occur during use in the application and/or in interaction with other components of the module. Because of that cell degradation effects are usually addressed in full-size module tests, for which also realistic voltage and current levels can be applied. To analyze the cell in small laminates as described in Section 6.2.2, the following tools can be used:

- IV curve measurement to check the electrical performance
- Electroluminescence to check for cracks or breakage
- Photoluminescence

Further analytical tools can be applied if the cell is removed from the laminate after the testing. These tools are covering all issues of semiconductor analytics. Since the investigation of all cell properties would be beyond the focus of this book, interested readers are recommended to have a look on the literature focusing on cell development and characterization.

There is no accelerated weathering testing procedure established for cells since they are generally seen to be intrinsically stable against influences of temperature, humidity, and irradiation in the intensities occurring in PV modules. If specific new substances in or on cells shall be tested, like cell coatings, it is recommended to use testing approaches similar to the ones described in Section 6.2.2.

6.2.4 Backsheet materials

Backsheet materials separate and protect the inner materials of PV modules from the ambience. They so have a safety and reliability-relevant function and are in direct contact with climatic loads. Key functions and properties backsheets have to ensure are the electrical safety and barrier function. They have to withstand the given microclimatic loads inside the module like elevated temperature and humidity, but also UV radiation from the front side passing between the cells and potential chemical loads, for example, from degradation products of the encapsulant. On the outer side, the rear side of the module, they are in contact with the ambience, including typical climatic loads and mechanical impacts (Fig. 6.6). It has to be mentioned that PV modules in open rack installations also have to stand a significant amount of UV radiation from

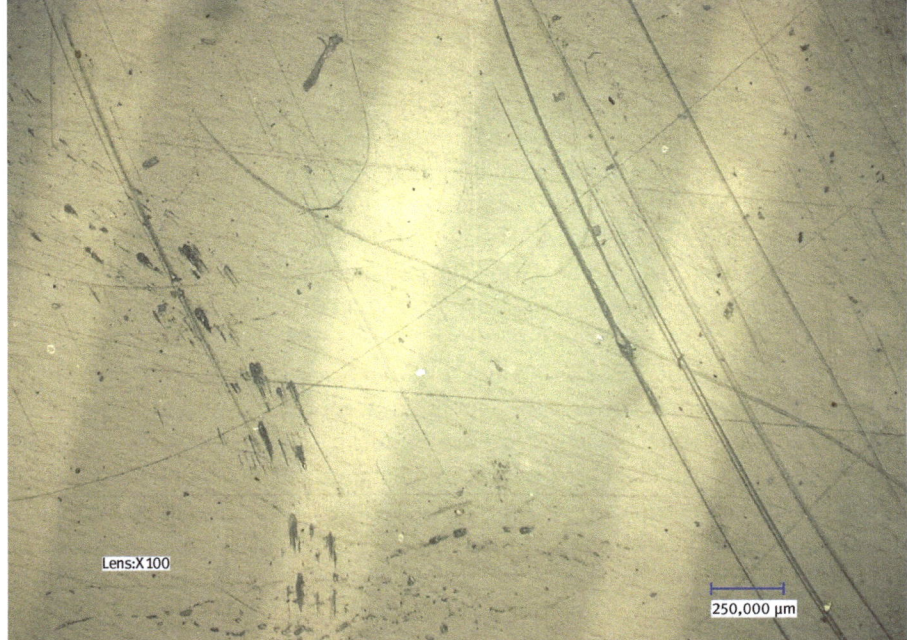

Lens:X 100

250,000 μm

Fig. 6.6: Scratches on the surface of a backsheet.

the rear side. Depending on the local situation, albedo radiation can sum up for up to 25–30% of G. Backsheets typically consist of a multilayer design with different materials, often including adhesive layers. Usually outer layers are produced by very weathering stable materials, like fluorinated polymers (e.g., PVF or PVDF) or highly stabilized PET, to make them more resistant against climatic loads. The core layer is mainly responsible for electrical insulation properties and mechanical properties, usually consisting of a material with lower stabilization to reduce costs like PET, PA, or polyolefins. Usually, the core layer clearly dominates the thickness of the total backsheet. The backsheets are usually produced, for example, by coextrusion of different materials or by lamination of different polymeric foils, connected with adhesives. Due to this complex structure of the backsheet, it is important to ensure that effects of all layers of the sheet are included in the tests. This inclusion requires awareness of the load levels in the microclimate for each layer in normal operation and also in the accelerated tests and it also requires that degradation analytics not only addresses properties of the complete backsheet but also checks for changes of the different layers as they can cause problems in future.

The following tools are typically used (depending on the available equipment and the specific purpose of the test) to analyze backsheet materials:

- Reflection measurement, for example, FT-IR spectroscopy is used to determine the optical properties of the material.
- Raman spectroscopy/microscopy to analyze the bondings of the material, so to gain information on the chemical composition. In confocal mode, this method can also deliver information with lateral resolution and also generate depth profiles, even through the outer layer of a multilayer laminate.
- Fluorescence measurements are mainly used to look for degradation products since these often contain double bondings that generate the fluorescence signal.
- DMA and tensile testing are used to analyze mechanical/elasticity properties.
- Tensile testing/peel testing is used to analyze the adhesion to encapsulation
- Electrical testing is used for breakthrough voltage.
- Permeation measurement is used to determine barrier properties.

All backsheets are, since they are made of polymeric materials, to a certain extent sensitive to weathering loads like temperature, humidity, and UV radiation. Some tests of backsheets can be performed on the sheet itself but for some potential failures or degradation effects, testing of laminates, as described in Section 6.2.2, is also necessary and recommended. Especially the electrical and insulation testing has to be performed on the backsheet alone. Nevertheless, degradation processes can be different if the material is exposed to tests without the influence of other module components. It is recommended to perform standard degradation tests, such as exposure to high temperatures, to UV radiation (from both sides is very important!), or to DH conditions, first, with the backsheet itself. Samples of these tests should be used to analyze the insulation properties after the tests and also the barrier properties. The barrier properties, which can be measured by a setup as presented in Fig. 6.2, are not a typical reliability parameter. The data is necessary to define appropriate testing conditions for modules and materials and should therefore be determined (if possible temperature dependent) directly when backsheets are tested.

Since backsheets comprise a number of interacting materials, they have different sensitivities toward climatic loads. It is of special importance to ensure that the testing conditions cover all relevant load conditions. It has to be mentioned again that the load conditions depend on the location and installation and operational conditions. Unfortunately, very serious issues occurred in the past affecting several GW of installed PV plants, which can be directly related to backsheet materials that have been comprehensively tested – but some combinations of loads have not been included. In the recent case, backsheets based on PA materials are affected, and it has been shown that the combination of chemical degradation caused by UV and temperature and humidity combined with mechanical tension causes cracks after some years of normal operation (Fig. 6.7). The effect can be seen in lab tests after relatively short testing times – if the right combination of tests is chosen (Fig. 6.8).

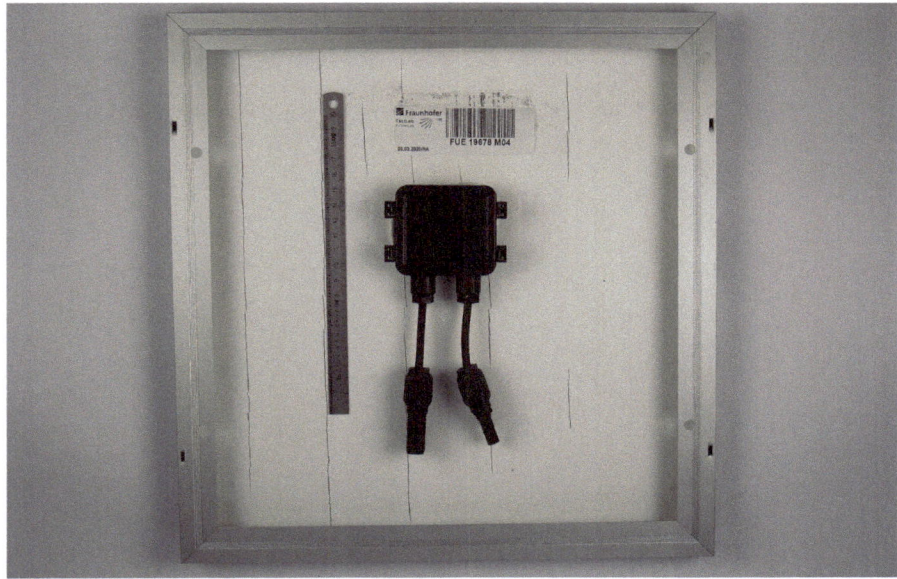

Fig. 6.7: Multiple cracks in the backsheet of minimodule after serial testing in DH conditions followed by two cycles of UV and temperature cycling testing.

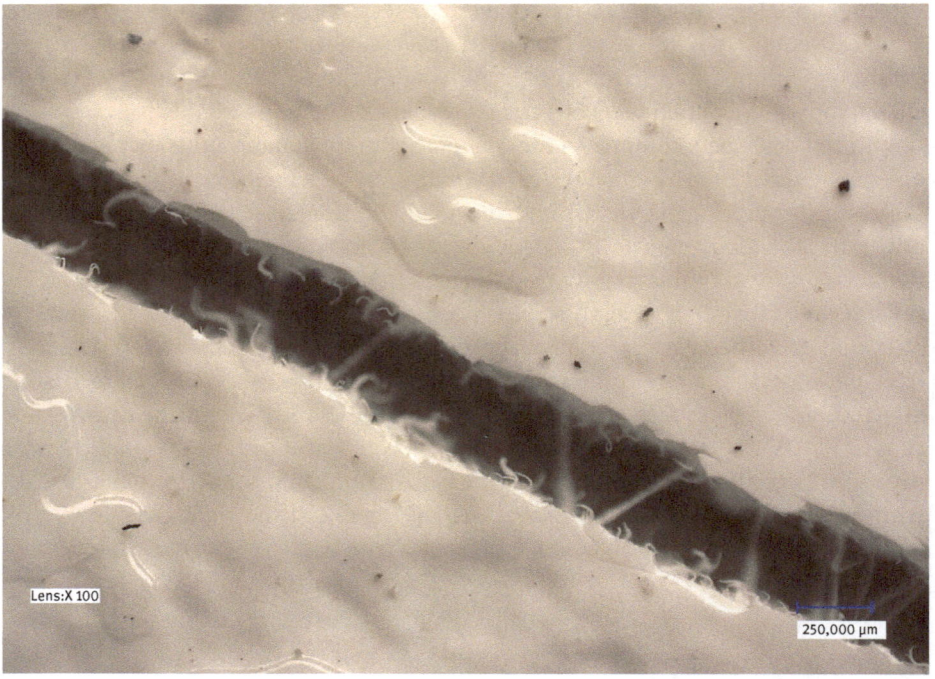

Fig. 6.8: Microscopic image of cracks in the backsheet showing also the used glass fiber in the material.

Karl-Anders Weiß

7 Reliability testing of modules

Since the photovoltaic (PV) module is the marketable product, its reliability is in the end requested by the end user and investors of PV plants. All testing of materials can be seen as a prerequisite to produce reliable modules but in the end reliability has to be proven on the module level. Also several degradation processes and effects require that all components of a module are available or are depending on the scale effects. This means that they cannot be properly tested with smaller samples or samples with reduced complexity, as also described in the sections on material testing in Chapter 6. Therefore, many tests to analyze reliability only make sense on the level of full-size modules.

7.1 Equipment

The general equipment that is used for reliability testing of PV modules is similar to the types of devices and systems described in the accelerated aging (Chapter 5) and reliability testing of materials (Chapter 6). As a matter of fact, the equipment needs to be bigger and more powerful in order to ensure that full-sized modules can be tested. This can be a challenge if, for example, irradiation, humidity, and temperature need to be controlled simultaneously. Especially equipment allowing fast and/or combined testing for such big samples like PV modules is also quite expensive and therefore not available at all labs.

Several pieces of special equipment for reliability testing of PV modules have been developed since the requirements of reliability testing in PV could not be addressed with standardized equipment used also in other industries. Such testing setups are, for example, solar simulators which allow a very high quality and stability of the spectrum and in parallel highest spatial uniformity for testing areas of 4×4 m^2 or even more. Other testing setups allow mechanical testing under controlled temperature and/or humidity conditions to simulate the typical winter loads for PV modules. Climatic cabinets enabling the exposure of several full-size PV modules to high-intensity UV irradiation and high humidities in parallel under controlled temperature conditions have specifically been designed to provoke chemical degradation processes of the polymeric components in the modules.

https://doi.org/10.1515/9783110685558-007

7.2 Procedures

Reliability testing procedures for PV modules address the reliability of the complete module and have to always take into account interactions of materials within the module and of the module with the ambience (in outdoor operation or in testing equipment). Often testing procedures are designed to address specific failure modes and/or for specific module types or designs to reduce complexity.

The development of a universal reliability testing procedure for all kinds of modules addressing all possible failure and degradation modes is not realistic, since the acceleration factors of processes depend on the materials and material combinations. In addition, there are different climatic conditions for PV modules installed in different regions and these conditions can provoke different processes. In the following, the general approach to design and perform reliability testing will be described and a special focus will be on climate-adapted testing since this becomes more and more important due to diversification of the PV market and applications.

7.2.1 General approach

The major difference between accelerated aging and reliability testing lies within the general approach to the testing procedure. In contrast to the load-based approach of accelerated aging tests, which apply one or multiple loads and investigate the effects, reliability testing aims at reproducing specific **degradation phenomena**, which tend to occur under operation. Therefore, it is even more important for the reliability testing of modules, compared with that of materials, to ensure that during tests **realistic conditions** occur. Especially the microclimatic conditions within the module are of high importance for the materials. This does not mean that the tests cannot use exceeding loads in comparison to the real operation for acceleration, but rather that the conditions have to be chosen in a way that no effects occur which are unrealistic for field exposure. It is always possible to "kill something by testing" if this is not taken into account. Still there is a high pressure from the market to make tests as short as possible to reduce costs and also to **keep up with the development cycles** of new products. The idea of this basic approach is to **increase load intensities** to **shorten testing times**. This is easiest by using higher temperatures – or in the case of PV use higher irradiation intensities, which in the end also increases temperatures. If tests are not carefully designed, it is highly probable that temperature levels are reached in samples that are far beyond operational temperatures. These temperatures can be in the range where additional processes are provoked, thereby causing fast degradation due to processes that would never occur in normal operation. Thus, such tests will generate results and perhaps even enable a rating of products, but the results would simply be meaningless for normal operation.

Some of the phenomena, which are reproduced in the modern state-of-the-art reliability testing, are potential-induced degradation (PID), snail tracks, or discoloration. A detailed description of these effects is given in literature [51–54]. In Fig. 7.1, a schematic description of a reliability testing procedure is shown. Mandatory tests for characterization, such as power measurements, visual inspections, insulation, and wet leakage tests (all described in standards, e.g. IEC 61215), electroluminescence, and – when more comprehensive characterization is required – Raman spectroscopic or fluorescence measurements should be performed after every test cycle. Additional characterizations can be performed after each aging sequence within the test cycle. Additional characterization is utilized for more detailed analysis when degradation effects are determined. Such test cycles enable to compare the reliability of different module types. Since they expose the specimens to relevant loads, changes in properties of the module and its materials can be identified at an early stage, if suitable characterization methods are used, often much earlier than they would be recognized by power or insulation measurements. It must be mentioned that most test cycles do not include all relevant loads, due to cost and/or time limitations. Here, a careful selection is necessary in order to include the most important loads and load combinations.

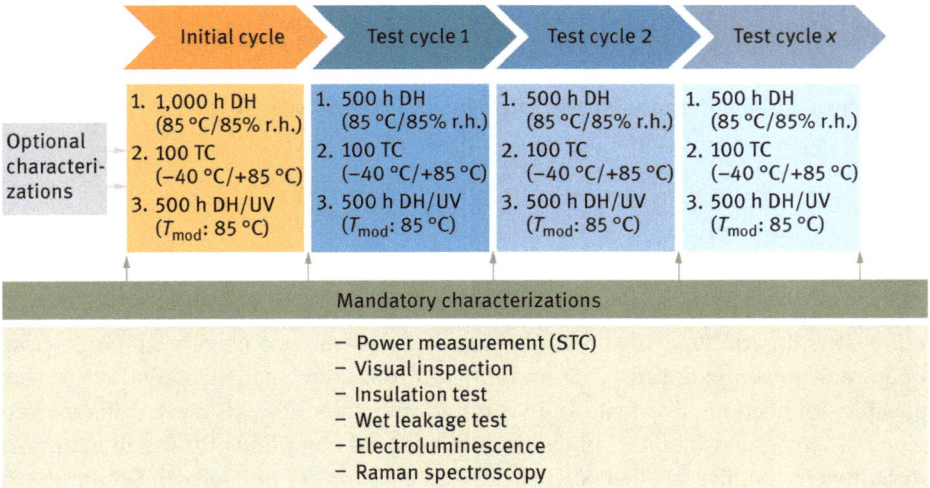

Fig. 7.1: Schematic description of a reliability testing procedure for PV modules. Specific tests are to be seen as examples.

The rating of loads and load collectives depend very much on the module design, the used materials, and the climatic conditions as well as the type of application (see Section 7.2.2). The sequence shown in Fig. 7.1, for example, does not include external mechanical loads, which would simulate snow loads or high electrical potential loads that can cause PID effects of PV modules. This cycle focuses on chemical and physical degradation effects caused by high temperatures, humidity, and UV irradiation, represented by damp heat and damp heat with UV tests. It further investigates the effects of internal mechanical stress, represented by temperature cycling tests. Such tests do not automatically deliver results with regard to expectable lifetime of the tested module types. Since local operational conditions and climatic loads have a strong influence on degradation effects, the time of service corresponding to one of the cycles shown in Fig. 7.1 depends on these local and operational conditions. It is also absolutely important to mention that such test sequences have to be adapted to each specific module type if real service lifetime testing is required. The rate dominating process, defining the acceleration factor, is specific for the material and material combination. A description of the design and application of service life tests would go beyond the topic of this book, especially since PV modules consist of a combination of different materials and material classes and have to bear a variety of loads and load conditions. A more detailed description of service life testing can be found in [55] or other publications focused on service life prediction.

If yields of PV modules over time are to be predicted, it is also important to analyze the behavior of modules under conditions, which do not only cause degradation of materials but can cause additional effects reducing the performance, such as high dust loads leading to soiling of the modules.

When reliability tests are performed with the aim to enable service life prediction calculations, a variation of intensities of the included loads is necessary. This is a prerequisite to determine acceleration factors (see Section 5.4) which are mandatory for the service life modeling, as described in Chapter 9.

7.2.2 Climate- or application-adapted testing

PV systems and modules are applied worldwide and also in all kinds of climatic conditions. In addition, the operational conditions can also be very different, from open-rack power plant setup, via on-roof or in-roof or in façade installations to very specific applications like floating PV or road-integrated PV. All these different versions of PV systems come along with different load conditions for the modules, as described in Chapter 4, especially with the calculations of humidity in Section 4.1.3. These conditions influence degradation and reliability and therefore have to be taken into account when meaningful reliability testing is required. Especially the ranges and the combinations of loads are important and can be quite different, as shown in Figs. 7.2 and 7.3.

There are sets of load data available for some established operational conditions, such as roof-integrated modules in Central Europe or open-rack utility size systems in northern Africa. These data sets also include microclimatic information like module temperature and can be used to adapt the tests according to the procedures described in Section 5.3 and in previous sections. For applications in locations where no data

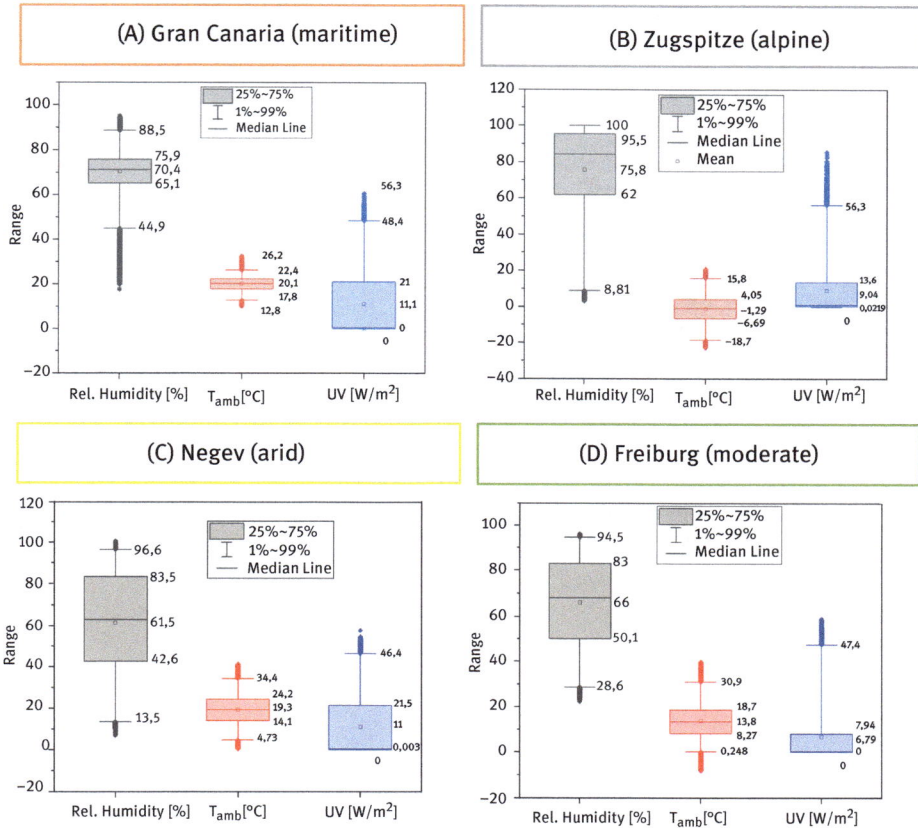

Fig. 7.2: Ranges of climatic loads (ambient data) of four locations with different climatic conditions: (A) maritime, (B) alpine, (C) arid, and (D) moderate. The box plots show not only the typical conditions but also the extrema.

are available, estimations for the load conditions need to be made. There are models using climatic data as input, as described in Section 4.1.2, which help to estimate the conditions.

It is important to analyze not only the climatic data but also the surrounding of the location to check if some additional loads have to be expected. Such loads can be caused by industrial exhaust gases or agricultural exhausts (ammonia – possible but rare) or also by corrosive substances coming from the sea close by or due to mining or very salty soil.

All these information on the specific site of interest has to be included in the definition of tests and test sequences since they influence reliability. The set of potential tests is the same for most situations but the applicable combination of tests and the intensity and dose levels (temperatures, UV dose, snow loads, . . .) depend on the specific site. Special loads as described earlier, such as corrosive loads or

Fig. 7.3: Average hourly irradiation (red, in W/m²) and relative humidity (blue, in %) of an arid (Negev desert, Israel) and an alpine (mountain Zugspitze, Germany) site showing very dry daytime hours with high irradiation in the desert and all day relatively high humidity levels on the mountain Zugspitze. The graphs show the measured data in 2014.

abrasive loads, can be neglected for most cases but are crucial for some and there-fore always have to be taken into consideration when deciding about reliability tests for PV modules.

Classification schemes for climatic conditions as described in Section 5.3 can make the development of adapted tests easier since they generalize some conditions for areas with comparable load conditions. It is anyway recommended to take the dif-ferent load levels and conditions into account when products are developed for the market. The size of the global PV market nowadays makes it possible to develop prod-ucts and tests optimized for different locations and applications. This way costs can be reduced with even positive effects on reliability.

Bengt Jäckel

8 PV module and component standardization

Standardization of photovoltaic (PV) modules and components is organized and regulated on the international level by the technical committee (TC) 82 "Solar photovoltaic energy systems" of the International Electrotechnical Commission (IEC).

The standards are under regular discussion to follow scientific and technical progress and industrial needs. Revisions occur and are published via corrigendum (if something went wrong in the publication process), amendments (additions, if technical modifications are not significant), or new editions (major changes to the standards – compliance testing requires generally full verification and testing).

8.1 General information

On national level exist the so-called mirror committees, which cover similar topics and organize the exchange between national and international levels by sending representatives to IEC meetings and transfer IEC standards to national standards. In many countries, IEC standards are translated into national standards, for example, European (EN) or German (DIN) standards. They use the same numbers, for example, IEC 61215-1:2016. 6xxxx numbers indicate IEC standards that are typically transferred via CENELEC (EN) into national standards, for example, the Deutsche Kommission Elektrotechnik (DKE) in Germany into a DIN. 5xxxx numbers indicate European grown standards. There is no IEC equivalent. If IEC adopts an EN standard, it gets a new IEC 6xxxxx number and the EN 5xxxx version will be withdrawn. The communication between EN and IEC is regulated by several agreements like the Frankfurt and Dresden agreements that define the interaction and communication of new standard proposals.

This chapter describes the situation at the mid of 2020. The latest versions of IEC standards are published by IEC and national bodies and can be found at https://webstore.iec.ch. Work in progress can be found on the IEC TC 82 website at https://www.iec.ch/dyn/www/f?p=103:7:0::::FSP_ORG_ID,FSP_LANG_ID:1276,25. Access to documents is limited and can be achieved by being a national representative of a national committee.

Acceptance of certification test reports between accredited certification bodies (CBs) is currently organized by IECEE (IEC System of Conformity Assessment Schemes for Electrotechnical Equipment and Components https://www.iecee.org/). The scheme is rarely used in PV and mainly for the type approval of major PV components (PV modules and inverters). Some years ago, a new system has been established to specifically cover the needs of renewable energies. The new scheme is called IECRE (IEC System for

https://doi.org/10.1515/9783110685558-008

Certification to Standards Relating to Equipment for use in Renewable Energy Application https://www.iecre.org/). The new scheme's mission is as follows:

> As a fundamental pillar enabling sustainable electricity generation and supply, IECRE is the internationally accepted conformity assessment system for all power plants producing, storing, or converting energy from renewable sources; which ensures that essential quality and safety standards are met, and as a consequence, reliable performance can be expected.

At the beginning, mainly conformity assessment was discussed and devolved while integrating IEC standards into the published Operational Documents that cover, for example, factory inspections for PV modules.

At the moment, it is under discussion to merge all IECEE activities from renewable energy to the new renewable energy systems.

Beyond the activities in IEC, there are other standardization bodies addressing other parts of the value chain, including International Standardization Organization (ISO) (https://www.iso.org), North American standards like ANSI/UL (https://www.ansi.org/; https://www.ul.com/), and more production-focused documents by SEMI (https://www.semi.org).

For ensuring reliable measurements and quality assessment, ISO standards are used. Test labs follow the accreditation procedures from ISO/IEC 17025 to test according to these standards. ISO/IEC 17065 adds further requirements for CBs. In general, most companies follow ISO/IEC 9001 for their quality management system. In recent years, other standards have been developed, to cover also environmental and health aspects (ISO 14001, 45001, 50001). In the PV industry, around 200 standards exist, excluding factory-focused procedures. Most of them are published within the IEC TC 82 committee. The documents cover all parts of a PV system, starting from materials, components (e.g., cables and connectors), PV modules, PV inverters, measurements procedures, and acceptance testing in the PV power plant. As this book is on PV modules, only those standards most relevant for modules will be considered. The rest of this chapter will go into more details of the PV standards structuring the content as follows:

1. Materials and component testing
2. Safety testing
3. Type approval
4. Extra testing procedures for PV modules
5. Measurement techniques
6. Power and energy yield

8.2 Materials and component testing

In the past decade, a lot of work was done establishing standards for materials and component testing. IEC 62788 series, for instance, is a series of standards covering backsheets, encapsulants, edge seals, and even front glass coatings. The intent of the documents is to prequalify materials for the use in a PV module with the aim to last 25+ years in the field in different climatic conditions. A just recently announced new project called IEC 63209-2 "Extended-stress testing of photovoltaic modules for risk analysis – Part 2: Durability characterization of polymeric component materials and packaging sets," is further aimed to bridge the gap between materials, coupons, and full PV module and can be considered as a work-in-progress document at the moment.

PV module manufacturers relay on several components they purchase besides materials. The main focus here is on electrical connections, including connectors (IEC 62852), cables (IEC 62930), diodes (IEC 62916), and PV module junction boxes (IEC 62790). Those documents typically cover the safety aspects of the component, and its intended use (clearance and creepage distances, flammability, and impact resistance of the used materials) is specified by the manufacturer (e.g. system voltage and currents under normal use).

Those materials and component tests do not replace PV module end product verification and testing. They, in general, neglect interactions between different materials, but they are helpful to generally prove the applicability of the material.

8.3 Safety testing – IEC 61730

Safety testing of PV modules is required by law in many countries and is covered by IEC 61730 series. The European low-voltage directive (LVD), for instance, calls out specifically the requirement of compliance regarding EN 61730 series (IEC 61730-series with European deviations including Annexe ZA and ZZ)[1] for CE marking, which is a requirement to bring products into the EU market. Other countries do have similar requirements stated in their national electric codes.

IEC 61730 series is divided into two parts:
- IEC 61730-1 photovoltaic (PV) module safety qualification – Part 1: requirements for construction
- IEC 61730-2 photovoltaic (PV) module safety qualification – Part 2: requirements for testing

Both parts were newly published in 2016 and are currently under revision via amendments to cover recently found problems. Both documents will be most likely published in 2021. Compliance against both parts is required by the LVD.

---

[1] ZA: Normative references to international publications with their corresponding European publications. ZZ: Relationship between this European standard and the safety objectives of Directive 2014/35/EU [2014 OJ L96] aimed to be covered.

Part 1 describes the general requirements that have to be met by materials and module design, for example, distances between life parts and the frame or properties of materials (overvoltage category, pollution degree, materials class, clearance, and creepage distances). The specific values are stated in tables 3 and 4 of the standard and are further dependent on the use case of the PV module (typically Class II, 1,500 V).

Part 2 describes tests, sequences, and pass/fail criteria for safety testing of PV modules. Similar to IEC 61215 series, it consists of characterization tests to measure the status or changes of the samples and accelerated weathering tests. Since several tests are identical to the ones in IEC 61215-2, tests in IEC 61730-2 are typically referenced to IEC 61215-2. As some of the sequences are very similar, it usually makes sense to perform the tests for IEC 61215 series and 61730 series for one module type in parallel. Furthermore, only the most important additional requirements and tests of IEC 61730 are described. The naming and numbering of tests in IEC 61730-2 uses the short form module safety test (MST) for each test. Several tests are included in IEC 61215 (module quality tests (MQT)) and additionally have an MST number, since they are part of IEC 61730. The major MST tests for modules are listed and commented:
- [MST 11] accessibility test;
- [MST 12] cut susceptibility test: test of the stability of modules against mechanical impact, for example, with tools, mainly relevant for backsheets;
- [MST 13] ground continuity test;
- [MST 14] impulse voltage test: test of electrical safety against high-voltage pulses;
- [MST 16] dielectric resistance test: also called high-voltage test, test of the electrical insulation properties of the complete module; procedure same as [MQT 02], except the use of increased voltages;
- [MST 17], also [MQT 15], wet leakage current test: test of the electrical insulation properties of the complete module under wet conditions, immersion of the module, and spraying of junction box;
- [MST 24] ignitability test;
- [MST 25], also [MQT 18.1], bypass diode thermal test: test of the stability and thermal behavior of bypass diodes under load;
- [MST 26] reverse current overload test: test of the module behavior under reverse current load;
- [MST 32] module breakage test: testing for mechanical integrity of the module after mechanical impact, performed on extra module;
- [MST 35] peel test: test of adhesion properties;
- [MST 36] lap shear strength test;
- [MST 42], also [MQT 14], robustness of terminations: test of the stability and adhesion of the junction box and the cables.

Qualification testing using the scheme of IEC with the standards IEC 61215 and IEC 61730 is an important pillar for international comparability of PV products. IEC 61730 series only proves product design compliance. It does not verify the ongoing production variations. Product variations are described in IEC 62915 (Retesting guideline) and production conformity is, at all, defined by certification bodies and guidance is given in the documentation of IECEE/IECRE. However, IEC 61730-2 states in Annex A tests for continous use for production verification.

8.4 Type approval testing – IEC 61215 series

In the 1990s, an international initiative was started to develop standards for the new PV industry to assure the quality and identify products of low material or production quality. This led to a first version of the type approval standard IEC 61215, crystalline silicon terrestrial PV modules – design qualification and type approval, which was published in 1993. The document was updated in 2005 and also thin-film technologies were introduced, covered by IEC 61646.

In 2016, a merged version was published, called IEC 61215 series. It combines the test sequences for crystalline c-Si and thin-film PV technologies into one document to handle all PV modules similarly. As each technology has some specifics, for example, dark storage or light-induced effects, each technology got a separate part to deal with just the differences.

Generally speaking, IEC 61215 series is applicable for PV modules which are used in general open-air climates, as defined in IEC 60721-2-1 on the ground, not for space applications, and also not for modules using concentrated sunlight (max 3X is accepted, if specified and accordingly tested).

It is very important to mention that passing the test sequences of IEC 61215 series proofs a specific technological level of module design, material, and production. It does not, however, do quality assurance, as it not includes independent inspections of the production. It further does not define a specific lifetime of modules that can be expected of the certified modules. The tests of the standard focus on the identification of specific known weaknesses or failure modes of modules. The focus lays more on failures, which can be called early failures or "infant mortality" of a module type, bearing the picture of the bathtub curve for failure rates in reliability engineering in mind. There is neither a scientific base for relating the tests to the life span of modules nor is the purpose of the standard to answer questions related to service life, even if many people in the industry expect this or even expect a correlation of IEC testing with given long-time warranties.

IEC 61215 series contains several test sequences. The sequences consist of characterization tests that measure or define a specific status of the module and also of weathering tests that expose the modules to defined loads in accelerated aging tests. Following are the most important or crucial tests described. A complete description of the tests and the sequences of IEC 61215 can be found in the currently valid version of the standard. A revised version of IEC 61215-series will be published in 2021 to include special requirements for Stabilization (MQT 19), PID testing (MQT 20) and flexible modules (MQT 22).

In the first version of the standard, just clause numbers were used to identify the tests. With the new edition, published 2016 MQT numbers are used . There are several significant changes with respect to pass/fail criteria introduced in the 2016 edition. Gates #1 and #2 define the levels the module must meet regarding performance (power output). Gate #1 verifies the rating of the module after stabilization (formally called preconditioning/light soaking), whereas Gate #2 checks for power loss within the subsequent testing sequences in the standard (basically the previous maximum allowed power loss, formally 8%, now 5%).

The following list highlights some key changes:

- visual inspection MQT 01: to define the visual status, criteria updated;
- performance under standard test conditions MQT 6.1 updated;
- nominal operating cell temperature (DUT (device under test) operated in U_{oc}) is replaced by NMOT (MQT 6.2) – nominal module operating temperature (DUT operated in P_{MPP} (maximum power point) mode);
- Bypass diode test has now two setups, one for thermal stability testing MQT 18.1 and the other for functionality MQT 18.2;
- stabilization of PV modules, formally light soaking/preconditioning MQT 19 is introduced with defined procedures in the technological subparts of IEC 61215-1-x.

The accelerated aging and weathering tests remain basically unchanged compared to the previous edition. Their main purpose, effects, and limitations are listed in the following points:

- [MQT 08] Outdoor exposure: exposure of modules to outdoors for at least 60 kWh/m^2 of sunlight, only a very low dose of sunlight, usually not enough to cause degradation or show weaknesses of materials or modules.
- [MQT 09] Hot spot test: testing of modules under illumination and partial shading of one cell to provoke hot spots and test stability and safety under these crucial conditions.
- [MQT 10] UV preconditioning (UV): exposure of modules to 15 kWh/m^2 UV light to provoke effects due to UV aging, for example, of polymeric materials; this UV dose only corresponds to a natural UV dose of 3 months in moderate climates like in Germany or even less in more sunny climates; the dose is usually not enough to show the weaknesses of materials, and the test is therefore only called preconditioning; UV preconditioning is followed by TC 50 test and humidity freeze (HF) test (see later);
- [MQT 11] Temperature cycling (TC): exposure of modules to 50 (TC 50) or 200 (TC 200), where a cycle runs from −40 °C to 85 °C to provoke effects due to internal mechanical stress, causing relevant and valuable effects but no scientific correlation to outdoor loads or specific service lifetime of modules,
- [MQT 12] HF testing: exposure of modules to 10 cycles from −40 °C to 85 °C and 85% relative humidity to provoke effects due to freezing of water or humidity which enter the modules or materials.

– [MQT 13] Damp heat testing: exposure of modules to 85% r.h. at 85 °C for 1,000 h to provoke effects caused by high humidity and/or high temperature, like hydrolysis of polymeric materials or corrosion of glazing or coatings, causing relevant and valuable effects but no scientific correlation to outdoor loads or specific service lifetime of modules.
– [MQT 16] Mechanical load testing (ML): exposure of modules to external MLs with the new differentiation between design load and test load. By default, a factor of 1.5 is used to determine the test load based on the manufacturer-defined design load. The minimum design load for normal applications is 1,600 Pa, resulting in a test load of ±2,400 Pa. Higher loads can be specified for upward and downward loads. Recently, a lower minimum criterion is discussed for applications behind the fence and for utility-scale power plants. The tests are performed at room temperature, which does not necessarily simulate realistic conditions for snow loads, since material properties are usually temperature dependent and cycle times do not simulate wind load conditions; strong gusts and wind introduced vibrations are not considered to be covered.
– [MQT 17] Hail impact test: exposure of modules to hail stones with defined size and speed, usually only causing effects if nontempered glass or polymers are used as front cover. Minimum hail size was increased to 25 mm.

In general, one can state that the tests of IEC 61215 series deliver valid and important results to identify modules of minor design quality and with clear problems of materials, material combinations, or production processes. Therefore, certification should be seen as absolutely necessary if the aim is quality and reliability. But it must also be mentioned that the full-type approval testing is only done with one small set of typically handpicked modules. The type approval, furthermore, does not cover different manufacturing sites and machinery. Theoretically, this is covered by certification and use of IEC 62915 (retesting requirements) but only in rare cases really applied. IEC 62915 is specifying most relevant modifications of a PV module type and states retesting requirements for IEC 61215 series and IEC 61730 series. The document is currently under revision to cover more topics and update certain requirements (e.g., definitions). It is also important to mention that certificates in general lose validity if module designs or materials are changed, which is often not easy for the customer to identify. The most important point to mention is that IEC testing has to be separated from service life testing, for which a totally different approach, including specific designs and materials, has to be chosen, for example, as described in Chapter 7. To sum up, testing according to the IEC standards should be seen as necessary but not sufficient to identify modules of high quality with proven reliability and durability.

Extensions of the tests are often asked for and executed. Such extensions are good to determine failures regarding one stressor and can benchmark different designs. To reduce the number of extended stress test sequences, standards like IEC 62892, 63209, and 63279 are worked on.

As more and more power plants are built in hot climates, IEC 63126 was just recently published that deals with the problem of high-temperature applications, such as desert or building-integrated PV modules. The standard defines extended temperature intervals to better cover the daily temperature variations.

8.5 Extra testing procedures for PV modules

Besides the type approval tests described earlier, other, also field-relevant, tests and standards were developed. This section highlights a few as they are quite relevant to the industry.

8.5.1 Potential-induced degradation (PID) – IEC 62804

Potential-induced degradation (PID) phenomena result from the serial interconnection of PV modules, and their severity depends on the system design (system voltage, grounding, inverter topology, and weather). PID sensitivity depends on cell technology, encapsulation materials, and system design. Depending on the module design and technology, different tests have to be applied. IEC 62804 series specifies such requirements. Part 1 deals with "classical" PID of crystalline c-Si modules, Part 1–1 deals with delamination phenomena, mainly observed in c-Si module, and Part 2 deals with observation of thin-film technologies.

8.5.2 Salt mist corrosion – IEC 61701

PV modules installed close to the sea will see besides normal climatic impact (sunlight, temperature, and humidity) in addition to salt mist. To cover that extra path of degradation, a salt mist corrosion test for PV modules was developed. IEC 61701 is based on procedures used also for other industries. It is based on IEC 60068-2-52, and uses a salt mist of 5% sodium chloride solution at various severities (SG 1–6). The main difference between the severities is the duration of the test. For SG 1, the test runs 2 h, whereas the duration is 56 days for SG 6. Each SG is complemented by cleaning and final measurements to determine power loss of the PV module as well to verify electrical safety.

8.5.3 Ammonia exposure – IEC 62716

Besides the installation of PV modules close to the sea, a lot of area is available on stable roofs. Depending on the livestock, certain gases are produced and can cause damage to the PV module. Here, mainly ammonia was found to cause potential issues. IEC 62716 was developed to cover that particular issue. It determines the resistance of PV modules to ammonia (NH_3). All tests included in the sequences, except the bypass diode functionality test, are fully described in IEC 61215-2 and IEC 61730-2. They are combined in this standard to provide means to evaluate possible faults caused in PV modules when operating under wet atmospheres having high concentration of dissolved ammonia (NH_3).

8.6 Measurement techniques

There are several standards available describing the application of measurement technologies or interpretation of data.

8.6.1 Characterization of PV devices – IEC 60904 series

Currently, the IEC 60904 series consists of more than 10 parts dealing with characterization of PV devices. Part 1 deals with the measurement of PV current–voltage characteristics, including requirements for multijunction and bifacial devices. Parts 2, 3, 4, 7, 8, 9, and 10 deal with reference devices, calibration, traceability, linearity, spectral responsivity, mismatch correction, and sun simulator requirements. Part 5 deals with the determination of equivalent cell temperature.

More documents within the series are under preparation dealing with, for example, light-soaking/stabilization of solar devices prior to measurements (similar to MQT 19 of IEC 61215-2).

8.6.2 Infrared thermography of PV modules under operation – IEC 62446-3

Thermographic imaging is a very powerful tool to look for PV module defects in large quantities (see also Section 3.2.12). It offers a nondestructive in operation test procedure to check for defects. IEC 62446-3 lists several defects and states possible causes and actions to be taken for PV modules. Defects that can be detected are, for example, broken/heating cells, defective bypass diodes, noncontacted module string, nonconnected modules, and operation mode of the string (U_{OC} vs P_{MPP}). Generally, infrared can further be used to check for defects in electrical cabinets like combiner boxes, inverters, or other switch gear.

As described earlier, the procedure is well established to check large-scale PV power plants under operation. Thermographic image can also be taken on modules powering the PV module either in forward or backward direction. Both can help to find different kinds of other defects under laboratory conditions.

8.6.3 Electroluminescence

Electroluminescence is a powerful tool to check for defects in PV modules. In general, it works for both c-Si and thin-film technologies; for further description, see Section 3.2.2. But it is established the most to characterize c-Si PV modules for cell cracks caused by various reasons. A detailed failure catalog that is generally accepted does not exist, yet. IEC 60904-13 lays down some fundamental requirements and IEC 63202-2 states some defects and root causes on cell level, good for sorting in solar cell production, but not deals with descriptions and long-term effects of cell cracks/defects on module/system level. A new proposal was published recently for EL-Images taken outdoors in large scale PV power plants (proposed number IEC 62446-4).

8.7 Power and energy yield

There are also standards available that address questions related to performance of modules in real operation as well as measures to rate the yield of modules or systems.

8.7.1 Energy rating – IEC 61853 series

The IEC 61853 series deals with fundamental measurements to determine/predict energy yield. Part 1 deals with radiance and temperature performance measurements and power rating based on standard laboratory conditions. Part 2 deals with spectral responsivity, incidence angle, and module operating temperature measurements (definition of NMOT).

Part 3 uses the results of Part 1 and Part 2 to establish energy ratings of PV modules and can be combined with Part 4, including standard reference climatic profiles.

8.7.2 Module performance in systems – IEC 61724 series

IEC 61724 basically looks into the performance of PV modules more from the systems point of view. The series consists of three parts, where Part 1 deals with energy monitoring, Part 2 with capacity evaluation methods, and Part 3 with energy estimation methods. Here, for example, issues like mismatch in strings or soiling are considered to correctly determine the maximum power of plants.

Ismail Kaaya

9 Degradation modeling and service life prediction

Photovoltaic (PV) modules are often considered as the most reliable elements in PV systems. However, the question of "how long will a PV module last when exposed in outdoor operating conditions" is still a challenge. The outdoor stress factors described in the previous chapters influence the **reliability of PV modules** after several years of operation. The most accurate way to evaluate the reliability of PV modules is to monitor their performance in real operation. However, since the lifetime of PV modules is expected to be over 25 years, it will require waiting a considerable and economically unrealistic amount of time. Therefore, degradation models are utilized to estimate the reliability of PV modules and lifetime in a shorter period. Indeed, evaluating the reliability of PV modules within a shorter time is crucial for all **PV stakeholders** since, on the one hand, PV manufacturers benefit from setting realistic warranties as described also in Chapter 2. Usually the commercial PV module datasheets indicate linear **warranties** in terms of performance degradation. More specifically, some manufacturers give linear performance/degradation warranties with multiple steps. Figure 9.1 shows an example of a two-step warranty with 2.5%/a performance degradation in the first year of operation and 0.78%/a throughout the 2nd to 25th year of operation.

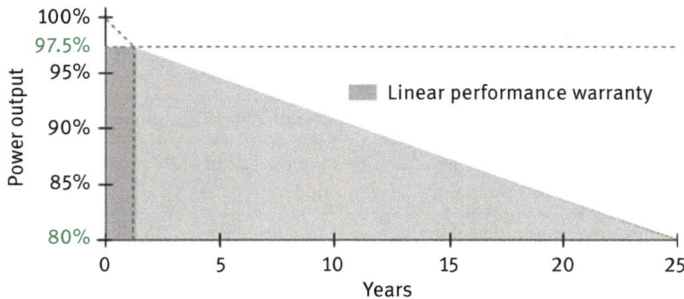

Fig. 9.1: Example of a linear performance warranty with an output decline of 2.5%/a at standard testing condition during the first year and 0.78%/a during the 2nd to 25th year.

It is not usually the case that the performance degradation will follow these linear trends [56]. Degradation models are therefore used to evaluate the nonlinearities in performance degradation [57]. This helps in setting more realistic performance warranties and expectations, hence, improving public trust in solar energy and accurate PV financial investments evaluations.

https://doi.org/10.1515/9783110685558-009

On the other hand, **degradation models** are utilized to evaluate the reliability of PV modules in **different operating climatic conditions** [58, 59]. Understanding how PV modules degrade in different climates is useful for researchers to design qualification standards and test procedures for different climates. Also researchers can improve the quality of PV modules' designs specifically for different climates [60] to increase the reliability of PV modules and systems.

Since some terminologies are interchangeably used or have no standard definition, they are defined in Tab. 9.1 as used in this chapter according to [59]. Especially failures are, in this chapter, only effects, which are related to gradual degradation. Spontaneous failures are not included in the models. A performance loss of 20% to define the end-of-life condition and so also a module failure is to be seen as example. The specific value depends on several factors, as also described in Chapter 2.

Tab. 9.1: Definition of the different terminologies used in this chapter.

Terminology	Definition
Performance degradation	Nonreversible gradual decrease in performance due to gradual degradation mechanisms
PV module failure	Performance decline by 20% of its initial power (nominal power)
Failure time	Time required for a PV module power to decay by −20% of the initial power
PV module reliability	Ability of a PV module to adequately perform until its failure period under the different operating conditions
Performance warranty period	Guarantee given by a manufacturer that a PV module performance decrease will not exceed a given limit in a specified period of time
Degradation model	Mathematical representation of the performance degradation over time
PV module reliability model	Time-dependent function that describes the evolution of PV modules' power over with increasing operation period
Degradation rate/ factor	Is a parameter that quantifies the magnitude of a PV module power decay of its initial power

9.1 Degradation models

Degradation models can either relate to electrical performance (P_{MPP}) or material deterioration. Although the two are correlated, however, degradation of a module component does not necessarily mean performance degradation. In most cases, a material might degrade without a substantial loss in module performance for a given period. Indeed, the yet unexplored challenge is a measure of what magnitude should a given material need to degrade to initiate a specific degradation in performance parameters.

A correlation of material degradation to performance degradation at different stress levels could be used to develop predictive models in a very short time.

Many methods have been proposed to predict and assess the long-term performance evolution of PV modules and systems. The methods can be grouped into three categories [67]:

- data-driven statistical methods,
- data-driven machine learning methods, and
- data-driven physical methods.

The term "data-driven" is used in all three methods due to one principle in common: fitting the available historical performance degradation data using regression or physical models to extract the degradation rates or to calibrate the physical models, respectively.

9.1.1 Data-driven statistical methods

Data-driven statistical methods (regression methods, autoregressive moving average models, etc.) are based on statistical analysis to calculate the trend of the PV performance time series and to translate the slope of the trend to an annual loss rate. This is the commonly applied approach in the PV community to estimate the degradation rates from historical performance data [69, 70].

9.1.2 Data-driven machine learning methods

Data-driven machine learning methods are based on machine learning algorithms (supervised, unsupervised, or reinforcement learning) to extract the degradation patterns and translate them into degradation rates. Like in other fields, machine learning methods are finding their application in photovoltaics. For example, researchers have applied machine learning methods to evaluate the degradation of PV modules [64] and to assess the long-term stability of emerging PV technologies, such as e.g. perovskites [65].

9.1.3 Data-driven physical methods

Data-driven physical models are developed based on the physical/chemical understandings and assumptions of specific degradation mechanisms. Degradation rates are estimated from physical models based on the input stress factors, shown by Kaaya et al. [58] and Subramaniyan et al. [65]. Generally, they are still only heuristic models which do not include the influence of all the intermediate degradation steps involved. In other words, the degradation kinetics of a specific degradation mechanism is modeled by assuming a rate-dominating process.

9.1.4 Hybrid modeling methods

Data-driven statistical and machine learning methods are believed to enable more accurate predictions, especially when enough data is available hence have promising potential for degradation rates evaluation. However, the main drawbacks of these two methods are the lack of a physical correlations or interpretation of the evaluated degradation rates because of their black box nature. On the contrary, physical models can provide a correlation of the estimated degradation rates to different degradation mechanisms or different climatic classifications as shown in [59]. However, their drawback are the different uncertainties associated with them. In [67] is shown that the uncertainties associated to physical models can originate from estimation of climatic stress factors (e.g. module temperature, relative humidity and UV irradiance) and also from degradation models simplifications. Hybrid models based on both, statistical and physical approaches, could provide a best solution for improving the prediction accuracy and retaining the physical interpretation.

According to the scope of the book, physical models are described as they provide a correlation of environmental stresses to PV modules' reliability. The discussion is also narrowed to degradation models of PV modules' power because this is the parameter used to set the performance warranty.

To begin with, it is important to understand how the physical degradation models are commonly applied in PV modules' reliability evaluation. The applications can be divided into three main categories:

A. First, the models are applied to analyze experimental observations. This is a common application for indoor accelerated aging experiments. During the testing of material degradation of PV modules, usually some material-dependent parameters such as the activation energy cannot be directly measured. They are, therefore, extracted using degradation models [67, 68]. This is achieved through fitting the models to the experimental data.

B. Second, when calibrated using the degradation history of a given location, the models are applied to predict the degradation rates for different locations. In addition, when global climatic data is available [58, 65], they are also applied to map the degradation rates in different climates [59].

C. Third, the calculated degradation rates are applied to forecast the lifetime of a given PV module.

9.2 Reliability models

The reliability models are time-dependent functions that describe the evolution of PV module power at a given time. Apart from the commonly used linear PV reliability model, several authors [58, 69, 70] have proposed reliability models to evaluate the nonlinearity of power degradation. Table 9.2 presents different PV module power reliability models. For each model, the failure time function is derived by considering failure as a 20% loss of the initial power.

Tab. 9.2: Different PV power degradation reliability functions with derived failure time.

Reliability model		Derived failure time		Parameter definition
Linear				$k(\text{year}^{-1})$ Degradation rate
$P(t) = P_0(1 - k.t)$	(9.1)	$t_f = \dfrac{0.2}{k}$	(9.2)	
Pan et al. [69]				β is a parameter associated with the material's natural lifetime
$\dfrac{P(t)}{P_0} = \exp(k \cdot t^\beta)$	(9.3)	$t_f = \left(\dfrac{\|\ln(0.8)\|}{k}\right)^{\beta-1}$	(9.4)	
Braisaz et al. [70]				B is not a defined parameter
$\dfrac{P(t)}{P_0} = \dfrac{1 + \exp(-B)}{1 + \exp(k \cdot t - B)}$	(9.5)	$t_f = \dfrac{\ln\left(1.25 \times \left(0.2 + e^{-B}\right)\right) + B}{k}$	(9.6)	
Kaaya et al. [58]				θ is a parameter associated with the material's natural lifetime μ is the shape parameter
$\dfrac{P(t)}{P_0} = 1 - \exp\left(-\left(\dfrac{\theta}{k \cdot t}\right)^\mu\right)$	(9.7)	$t_f = \dfrac{\theta}{k \cdot [|\ln(0.2)|]^{\mu-1}}$	(9.8)	

In all the reliability models (9.1), (9.3), (9.5), and (9.7), parameter k is the degradation rate/factor that relates the PV module life span to environmental stresses. Therefore, parameter k is a function of stress factors. Before discussing the functionalities of parameter k, a simple assessment of the above-presented reliability models is described.

Assuming a degradation factor $k = 0.008$ year^{-1} (0.8%/year) which corresponds to 25 years PV module lifetime ($t_f = 25$ years) using a linear reliability model. It is possible to evaluate the unknown parameters of eqs. (9.4), (9.6), and (9.8) as $\beta = 1.034$, $\mu = 0.30$ (assuming, $\theta = 1$), and $B = -10$. However, one should note that the parameter B of the Braisaz model has no solution at these values ($k = 0.008$ year^{-1}, and $t_f = 25$ years) as shown in Fig. 9.2. Therefore, $B = -10$ was used as an approximated value close to the solution.

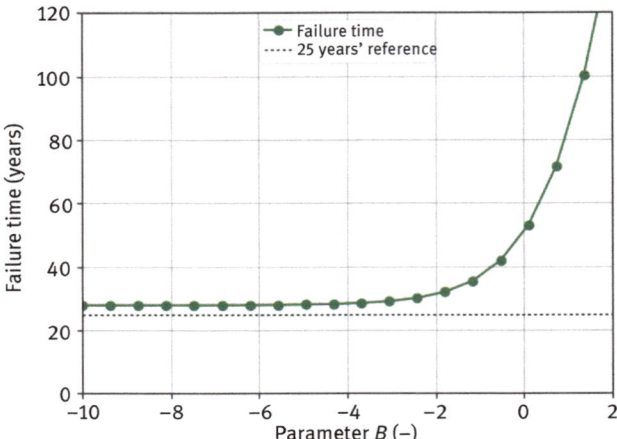

Fig. 9.2: Evaluated failure time for different values of parameter B of Braisaz model when $k = 0.008$ year^{-1}. It is seen that at this value of k, there is no solution for the failure time of 25 years.

Now that the unknown parameters are obtained, it is important to assess how they are interpreted in terms of reliability analysis. To do this, the models are plotted on the same axis as shown in Fig. 9.3. The key point to note here is that, although the models are evaluated using the same degradation rate and at the same failure time, it can be seen that they follow different trends. Indeed, this is important because it affects the evaluated energy yield. The energy yield is the integral power, represented in the graph by the areas under the given curves in Fig. 9.3. The overall energy yield is affected by the degradation shape despite a similar lifetime. Figure 9.3 demonstrates this effect, where the calculated relative energy yield is compared according to the different reliability models given at 25 years' lifetime. It can be seen that despite the similar degradation factors and module lifetime, the calculated yield varies accordingly among the models, which is an effect of the different degradation trends.

Using this analysis, the following can be summarized:

-   In reliability analysis, an assessment of the degradation trends is also a crucial part, since this affects the energy yield calculation, hence, the financial figures.
-   Due to the nonlinearity in degradation [56], care must not only be taken to select a model that takes into account the nonlinearity effect but also the model that takes into account the different degradation trends. By comparing the three nonlinear models of Pan, Braisaz, and Kaaya (see Tab. 9.2), it can be seen that the latter takes into account the degradation curves' shape effect. The model allows optimizing different degradation trends for same values of k and t_f by changing the parameters μ and θ as shown in Fig. 9.4.

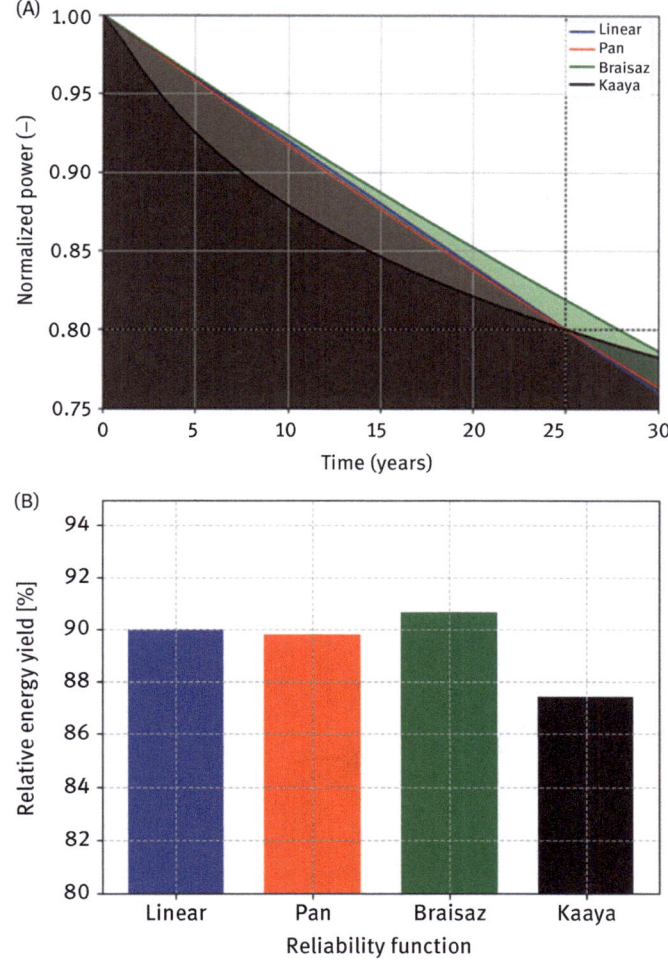

Fig. 9.3: (A) Evolution of power using different reliability models at similar degradation factors.
(B) Evaluated relative yield at 25 years' lifetime for the different reliability models.

(A)

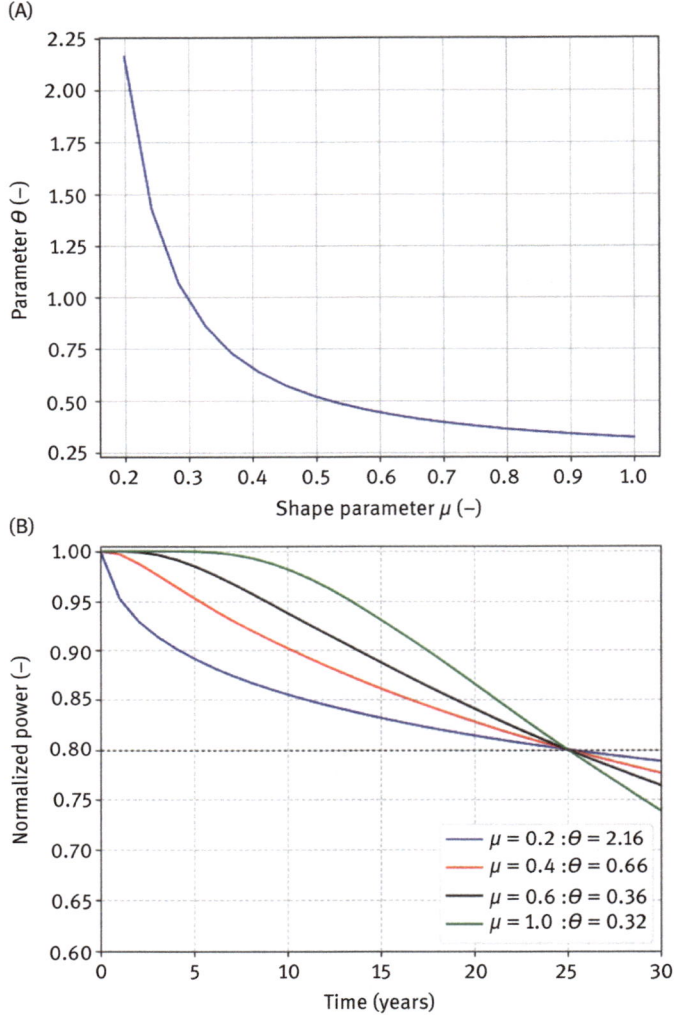

(B)

Fig. 9.4: (A) Change of parameter θ with different values of the shape parameter μ. Figure plotted using eq. (9.8) with $k = 0.008$ year^{-1} and $t_f = 25$ years. (B) Simulated degradation trends for different pairs of μ and θ.

9.3 Degradation rate/factor models

For well-known PV module degradation mechanisms, several degradation rate models are available [61]. All these models are based on the principle of quantifying the effects of climatic stress factors to the electrical parameters based on physical/chemical assumptions.

9.3.1 Modeling with indoor testing data

For indoor accelerated aging, the developed models are applied to analyze experimental observation of specific degradation mechanisms using controlled stress factors. For example, to analyze the degradation mechanism of PV modules due to temperature and humidity, the so-called damp heat (DH) test [67] is used. Although different models are proposed to evaluate the degradation mechanisms due to temperature and relative humidity [61], the most commonly used model is the Peck's model [72], expressed as

$$k = A \exp\left(\frac{-E_a}{k_B \cdot T}\right) \times (\text{RH})^n \qquad (9.9)$$

where E_a (eV) is the activation energy of power degradation, n is a parameter to quantify the effect of relative humidity to power degradation, A is the exponential constant, and k_B is the Boltzmann constant [8.62×10^{-5} eV/K].

By substituting eq. (9.9) into the reliability model (9.7), eq. (9.10) is derived. Using the DH degradation data of two modules from different manufacturers (named module A and module B as shown in Tab. 9.3), and eq. (9.10), it is demonstrated how degradation models are applied to analyze the experimental observations:

$$\frac{P(t)}{P_0} = 1 - \exp\left(-\left(\frac{\theta}{A \exp\left(\frac{-E_a}{k_B \cdot T}\right) \times (\text{RH})^n \cdot t}\right)^\mu\right) \qquad (9.10)$$

Tab. 9.3: Damp heat testing time and measured power in percentage of the initial power for two modules A and B. DH85/85 means the temperatures are kept constant at 85 °C and the relative humidity at 85%.

DH 85/85 Testing time (h)	Module A Power (%)	Module B Power (%)
0	100.0	100.0
500	99.9	100.0
1,000	99.4	99.5
1,500	98.7	99.1
2,000	98.7	99.1
2,500	96.0	98.3
3,000	93.5	97.8
3,500	90.7	95.3
4,000	89.8	94.4

Equation (9.10) is fitted to the experimental data of the two modules to extract the model parameters presented in Tab. 9.4. During parameter estimation, the optimization was done using a nonlinear least squares method. This method is inbuilt and available in most simulation software packages. Also to note is that, during the optimization, it is necessary to set some boundary conditions of the estimated parameters. The ranges of different parameters are available in different publications. To calibrate the data of the two modules, the following boundary conditions were used:

$$0.1 \leq \theta \leq 1.0$$

$$0.1 \leq \mu \leq 1.0$$

$$10 \leq A \leq 2e4$$

$$0.5 \leq E_a \leq 1.0$$

$$1.5 \leq n \leq 2.0$$

However, it is important to note that these parameters are very material or technology dependent. The residual standard deviation error (SD_{res}), which determines the quality of the fit, is evaluated as follows:

$$SD_{res} = \sqrt{\frac{\sum (Y_m - Y_f)^2}{n - 2}} \tag{9.11}$$

$$Residual = Y_m - Y_f \tag{9.12}$$

where Y_m is the measured value, Y_f is the fitted value, and n are the number of points.

Figure 9.5 shows the model fit to the two measured data sets of modules A and B and Tab. 9.4 lists the corresponding extracted model parameters for the two modules.

Tab. 9.4: Extracted model parameters from the equations fitted to the data set of two modules A and B. Similar boundary conditions were applied during optimization.

Module	Extracted model parameter at fit					SD_{res}
	θ	μ	A	$E_a(eV)$	n	
A	0.66	0.77	7,656.95	0.809	1.69	0.58
B	0.66	0.77	7,656.95	0.819	1.69	0.43

In this example, the model is used to extract material-dependent parameters such as the activation energy E_a. By comparing the extracted values of the two modules A and B, the dependency of the activation energy on the PV module degradation can be analyzed. From this example, it can be seen that with an activation energy of

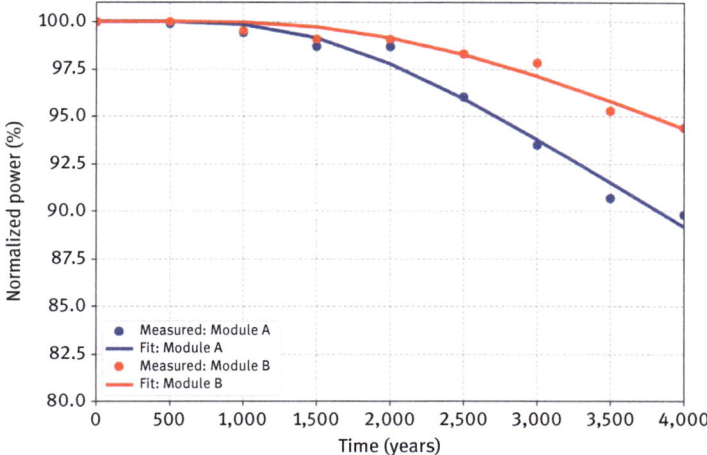

Fig. 9.5: Model fit on experimental damp heat (DH85/85) testing data of the two modules A (in blue) and B (in red).

0.809 eV, module A lost 10.2% after 4,000 h of testing, and at the activation energy of 0.819 eV module B lost only 5.6% of the initial power at the similar testing time and conditions.

To clearly understand what this means, $E_{(aref)} = 0.809$ eV, $T_{ref} = 85$ °C, and $t_{ref} = 4,000$ h are considered as the reference activation energy, temperature, and time, respectively, required to have a performance degradation of 10.2% for module A. Using eq. (9.9), it is possible to evaluate the time (t_i) required to achieve a 10.2% power loss at a given temperature T_i for different activation energy values $E_{(ai)}$, that is,

$$t\left(E_{a_i}, T_i\right) = t_{ref} \cdot \frac{k_{ref}}{k_i} = t_{ref} \cdot \frac{\exp\left(-\dfrac{E_{a_{ref}}}{k_B T_{ref}}\right)}{\exp\left(-\dfrac{E_{a_i}}{k_B T_i}\right)} \tag{9.13}$$

which can be simplified as

$$t\left(E_{a_i}, T_i\right) = t_{ref} \cdot \frac{k_{ref}}{k_i} = t_{ref} \cdot \exp\left(\frac{1}{k_B}\left(\frac{E_{a_i}}{T_i} - \frac{E_{a_{ref}}}{T_{ref}}\right)\right) \tag{9.14}$$

It should be noted that in the derivation of expression (9.14), the relative humidity and other parameters are kept constant. Using this expression, the required testing time to reach a degradation of 10.2% was evaluated as shown in Fig. 9.6.

Figure 9.6 shows the required testing time for different activation energy values (e.g., 0.788 to 0.830 eV) as well as for different testing temperatures 75 °C (green), 85 °C (blue) and 95 °C (red). The variation of testing time with temperature and

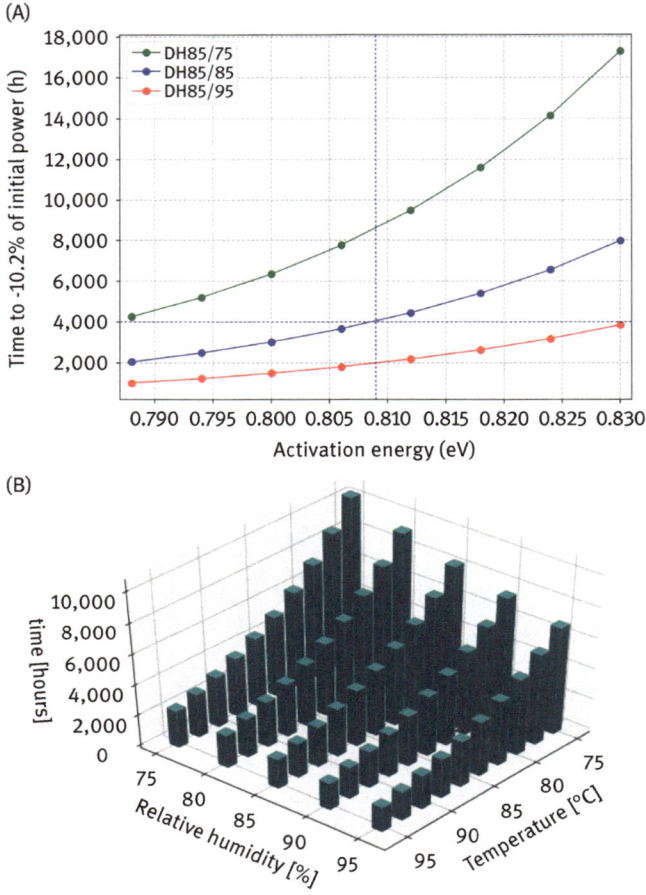

Fig. 9.6: (A) Required testing time for different values of activation energies and temperature. The blue dotted lines indicate the reference time and activation energy point. (B) Required testing time for different temperature–relative humidity values.

activation energy is clearly visible. For example, for the activation energy of 0.788 eV it requires only half of the testing time at 95 °C to have the same degradation as at 85 °C and the 2.1-fold testing time at 75 °C to have the same degradation compard to the reference at 85 °C. Of course, according to the nature of the graph, the required testing time will vary at different activation energies. Another important observation is on the sensitivity to the activation energy, for example, an increase in the activation energy by only 0.01 eV might lead to almost doubled necessary testing times. More-over, this also depends on the temperature, that is, for lower temperatures the sensi-tivity is higher compared to higher temperatures, as can be seen in the figure (e.g., for 75 °C and 95 °C). So, for a meaningful application of the models in relation to the experimental data, the activation energy sensitivity must be taken into account. That is, the activation energy should be varried in small range. Indeed, this is also evident in the experimental data where a difference in activation energy of only 0.01 eV between module A and B led to 45% relative difference in power after the same exposure time.

Lastly, the contribution of relative humidity should not be ignored as it is also an influencing factor. To evaluate the relative humidity dependence, the expression in eq. (9.12) can be rederived as follows:

$$t(\text{RH}_i T_i) = t_{\text{ref}} \cdot \frac{k_{\text{ref}}}{k_i} = t_{\text{ref}} \cdot \frac{\text{RH}_{\text{ref}}}{\text{RH}_i} \exp\left(\frac{E_{\text{ref}}}{k_B}\left(\frac{1}{T_i} - \frac{1}{T_{\text{ref}}}\right)\right) \qquad (9.15)$$

where RH_{ref} is the reference relative humidity used during model calibration and RH_i is the varying relative humidity.

Using expression (9.15) allows evaluating the required measuring time for different temperature–relative humidity combinations as shown in Fig. 9.6. Indeed, such analyses are used during the design of experiments for accelerated aging tests as described in Chapters 6 and 7.

9.3.2 Modeling with outdoor exposure data

So far, reliability models as well as degradation rate models have been introduced with a case study of their applications in experimental observation analysis. It is also noted that the analysis focused on degradation from indoor aging test where the degradation rate model evaluates temperature and humidity effects. In the following examples, the discussion or analysis focuses on degradation models developed for outdoor applications. The main differences between indoor aging tests and outdoor or real-life operation are:
- for indoor testing, the stress factors are controlled;
- this is not the case for outdoor with stochastic operating conditions;
- during indoor testing, specific conditions are applied to induce specific degradation mechanisms;
- in outdoor operations, the modules are exposed to numerous stress factors causing numerous degradation mechanisms at the same time.

From these differences, it is clear that correlating indoor testing conditions with outdoor conditions is a big challenge. This applies also to developing models for outdoor degradation evaluation. However, a few authors [58, 65] have developed models for outdoor applications.

9.3.3 Multiload models applied to location-specific loads

These models are developed to quantify the effects of combined climatic stresses on performance degradation. The models are calibrated using degradation data as well as climatic variables of a known location and are then applied to predict the degradation rates using climatic variables of other locations.

For example, in [65] an approach for PV module degradation rate modeling that includes the effects of module-specific stresses (module static temperature, module cyclic temperature) and other environmental stresses (ambient RH and plane of array UV radiation) was proposed as follows:

$$k(T, \Delta T, \text{UV}, \text{RH}) = \beta_0 \cdot \exp\left(\frac{-\beta_1}{k_B \cdot T_{max}}\right) \cdot (\Delta T_{daily})^{\beta_2} \cdot (\text{UV}_{daily})^{\beta_3} \cdot (\text{RH}_{daily})^{\beta_4} \qquad (9.16)$$

The parameters of the model are described in Tab. 9.5.

Tab. 9.5: Definition of model parameters.

Parameter	Definition
k (8.62×10^{-5} eV/K)	Boltzmann constant
T_{max} (K)	Daily maximum temperature of module
ΔT_{daily} (K)	Daily temperature delta of module
UV_{daily} (W/m^2)	Daily daytime average UV irradiance
RH_{daily} (%)	Daily average RH
β_0 (s^{-1})	Frequency factor
β_1 (eV)	Activation energy
β_2, β_3, and β_4	Model parameters that measure the effect of cyclic temperature, UV radiation, and RH, respectively

The model was calibrated on degradation data of monocrystalline PV modules. The calibrated model was applied to predict the degradation rates in four different regions with different climatic classification as presented in Tab. 9.6. The authors predicted strong degradation in Tampa, Florida, USA, with hot and humid climate.

Another model to quantify the effect of combined climatic stresses is proposed in [58]. In this approach, degradation rate models are proposed for specific degradation mechanisms/processes based on the applied climatic stresses. A combined/total degradation rate model was derived from the specific rate models as follows:

$$k_T = A_N \cdot \prod_{i=1}^{n} (1 + k_i) - 1 \qquad (9.17)$$

Tab. 9.6: Percentage power degradation per year for analyzed Si modules at various locations. Table from [65].

Region	Classification	Estimated degradation rate (%/year)	Lower 95% CI	Upper 95% CI
Phoenix	Hot and dry	1.50	0.90	1.92
Westchester	Cold	0.80	0.63	0.95
Tampa	Hot and humid	1.76	1.48	2.04
Albuquerque	Semiarid	1.17	0.63	1.47

where k_T is the total degradation rate (%/a), A_N is the normalization constant of the physical quantities, n is the total number of degradation processes, and k_i is the degradation rate of the ith process.

The three main degradation mechanisms were assumed as hydrolysis, photodegradation and thermomechanical degradation; hence, the total degradation rate due to these three mechanisms is expressed as follows:

$$k_T = A_N \cdot (1 + k_H)(1 + k_{PD})(1 + k_{Tm}) - 1 \tag{9.18}$$

where k_H, k_{PD}, and k_{Tm} are the degradation rates for hydrolysis, photodegradation, and thermomechanical mechanisms, respectively.

The degradation rates for specific mechanisms were evaluated as functions of climatic stresses as follows:

$$k_H(T, RH) = A_H \cdot \exp\left(\frac{-E_{aH}}{k_B T}\right) \times (RH)^n \tag{9.19}$$

$$k_{PD}(UV, T, RH) = A_P \cdot UV^y (1 + RH^n) \cdot \exp\left(\frac{-E_{aH}}{k_B T}\right) \tag{9.20}$$

$$k_{Tm}(\Delta T, T_{max}) = A_T (\Delta T + 273)^x \cdot C_r \cdot \exp\left(-\frac{E_{aT}}{k_B \cdot T_{max}}\right) \tag{9.21}$$

The parameters of the model are as in Tab. 9.7.

Tab. 9.7: Definition of model parameters.

Parameter	Definition
k_B (8.62×10^{-5} eV/K)	Boltzmann's constant
T_{max} (K)	Annual average temperature of module
T_{max} (K)	Annual average maximum temperature of module

Tab. 9.7 (continued)

Parameter	Definition
ΔT (K)	Annual cyclic temperature of module
UV (kWh/m^2)	Annual average ultraviolet irradiance dose
RH(%)	Annual average relative humidity
C_r (cycles/a)	Annual temperature cycling frequency
A_H (a^{-1}) A_P [(kWh/m^2/a)$^{-1}$] and A_T (cycles^{-1})	Exponential coefficients for hydrolysis, photodegradation, and thermomechanical rates, respectively
E_{aH}, E_{aP}, and E_{aT} (eV)	Activation energies for hydrolysis, photodegradation, and thermomechanical mechanisms, respectively
n, y, and x	Model parameters that measure the effect of RH, UV radiation, and cyclic temperature, respectively

The model was calibrated using degradation data of monocrystalline PV modules installed in Gran Canaria, Spain (arid–maritime climate), and validated using performance data of the same module type installed in the Negev Desert, Israel (arid climate), and on the mountain Zugspitze, Germany (alpine climate). The estimated degradation rates as well the climatic classifications of these three regions are described in Tab. 9.8. Figure 9.7 shows the climatic stresses (module temperature, relative humidity, and UV irradiance) of the three locations. The figures are generated based on monitored data of 2013. The module temperatures correspond to the measured temperatures of the monocrystalline silicon modules installed in these three locations.

Tab. 9.8: Climatic classification as well as the estimated degradation rates in the three regions.

Region	Classification	Estimated degradation rate (%/a)
Negev, Israel	Arid	0.74
Gran Canaria, Spain	Maritime (oceanic)	0.50
Zugspitze, Germany	Alpine (cold)	0.30

The authors predicted stronger degradation in Negev with hot and humid climatic condition which includes the predictions from Kaaya et al. [58].

So far, the above-presented examples have shown how the degradation models are applied as predictive models. The following example demonstrates how the degradation models are applied for a sensitivity analysis of a given module based on different climatic stress combinations.

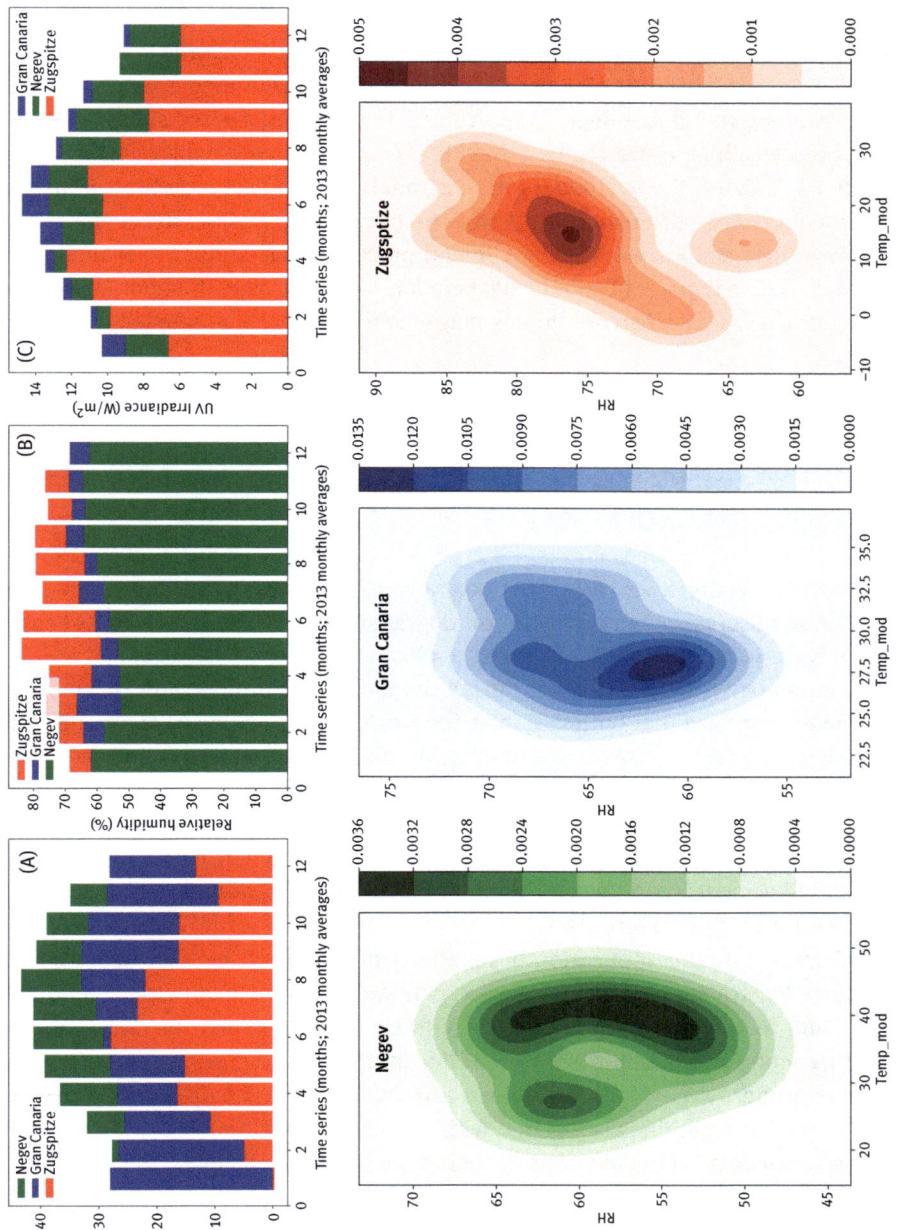

Fig. 9.7: (A) Monthly average temperatures; (B) monthly average relative humidity; (C) monthly average UV irradiance that a module experience in the three regions: Negev, Israel (green); Gran Canaria, Spain (blue); and Zugspitze, Germany (red) (Note: Barplots should be read starting from zero for all the locations). Bottom: annual temperature–relative humidity distributions for Negev, Gran Canaria, and Zugspitze in 2013.

Given the climatic stresses, the calibrated degradation rate models can be applied to analyze the sensitivity of a given PV module to the climatic conditions in different locations. For example, in [59], the degradation rate model (9.17) was applied to evaluate the global degradation rate of the monocrystalline silicon module using processed climatic stresses from the ERA5 climate reanalysis dataset [73].

Figure 9.8 shows the evaluated total degradation map based on the calibrated data of the monocrystalline modules. From the figure, it is possible to see the zones with stronger degradation worldwide. A similar procedure can be used to evaluate the effects and relative impacts of specific degradation mechanisms in different areas; this understanding is useful when developing new materials that can be adapted to specific climates.

9.4 Application of degradation models for long-term performance forecast

In the previous sections, we have described how degradation models are applied for experimental observation analysis and for degradation predictions as well as sensitivity analysis. It can be seen that in both applications, performance data as well as climatic data are required especially during the model calibration step. However, it is not usually the case that both variables (performance and climate) are monitored at the same time and in reasonable quality. In many situations, especially in commercial PV installations, only performance measurements (e.g., power) are available. And usually the system owners are interested to forecast the performance evolutions over the years to aid them in planning operation and maintenance activities. In this case, reliability models are used to fit the performance degradation data and extrapolate it for long-term forecasts.

To demonstrate the application of reliability models for long-term degradation forecasts, a PV system data set is used to calibrate the four mentioned reliability models. Due to the stochastic nature of outdoor measurements, before applying forecasting models, it is important to first decompose the time series data. Decomposition is primarily used for time series analysis, and as an analysis tool it can be used to reduce uncertainties associated with model calibration.

Time series data is characterized by four major components: level, trend, seasonality, and noise. Two approaches are used to model the effects of these components: additive (linear) and multiplicative (nonlinear) [74]. In this example, a multiplicative model is used. The model suggests that the components are multiplied together as follows:

$$y(t) = \text{Level} \times \text{Trend} \times \text{Seasonality} \times \text{Noise} \qquad (9.22)$$

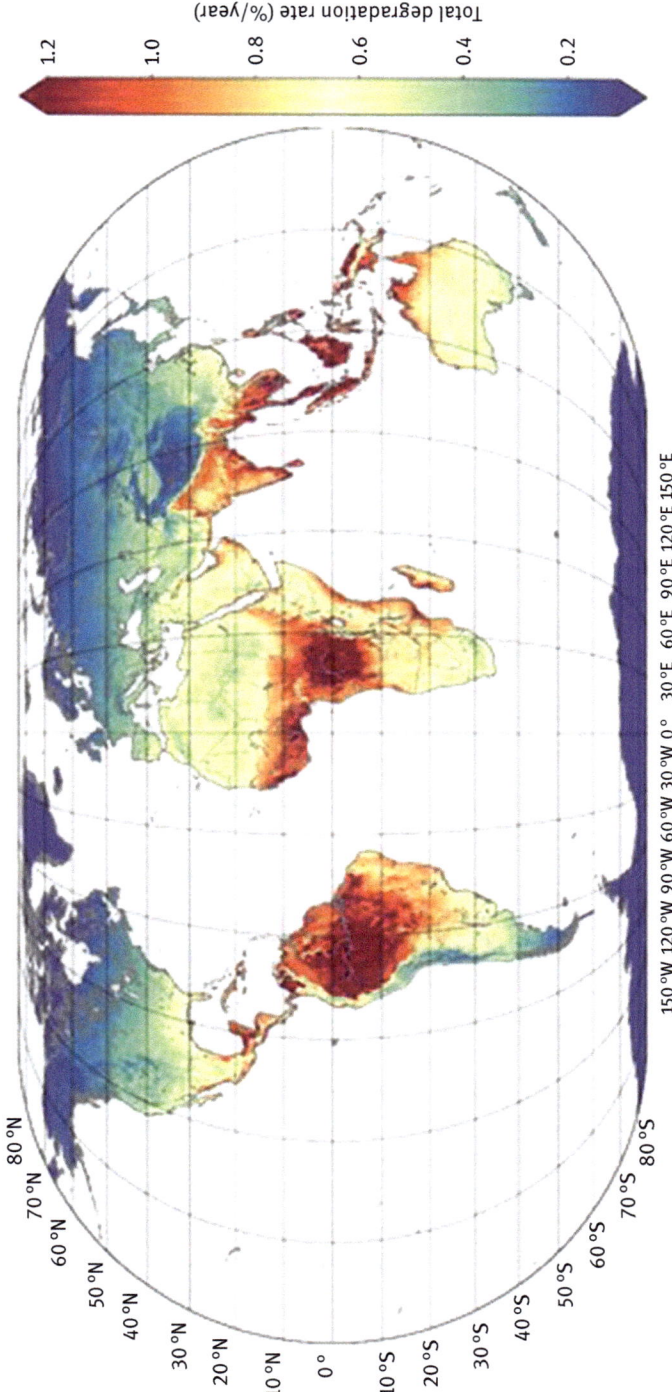

Fig. 9.8: Total degradation rates for a specific monocrystalline silicon PV module using the Kaaya degradation rate model. Climate data from ERA5 for the average between 2016 and 2018. Figure reprinted from [59].

The function uses a moving average method to extract a degradation trend in the time series data. The averaging depends on the required resolution (weekly, monthly, or yearly) of one's interest, and it determines the extracted degradation trend in this case.

In outdoor conditions, seasonal and other different effects can reduce the performance of a PV module or system. Therefore, it is crucial to choose a good averaging temporal interval that can separate reversible from nonreversible performance losses. (Degradation is taken as a nonreversible but gradual performance loss.) Figure 9.9 shows the multiplicative decomposition of the studied PV systems data with the four components: observed, trend, seasonal, and residuals.

The reliability models are calibrated on the extracted degradation trend; the model parameters as well as the degradation rates derived by calibration are shown in Tab. 9.9.

Tab. 9.9: Extracted degradation rates as well as model parameters during calibration. The failure time is calculated using the derived failure functions in Tab. 9.2, and corresponds to 20% loss of the initial power.

Reliability model	$k(a^{-1})$	Model parameter				SD_{res}	Failure time (a)
		β	B	θ	μ		
Linear	0.0051	–	–	–	–	0.011	39.6
Pan	0.0002	2.3	–	–	–	0.004	12.7
Braisaz	0.4320	–	6.2	–	–	0.005	11.1
Kaaya	0.0099	–	–	0.3	0.8	0.004	17.0

Figure 9.9 shows the long-term degradation forecast using the four reliability models. Even though all the models are calibrated using the same dataset, the forecasted long-term degradation varies significantly among the four models. Indeed, this is a strong point to consider when using the models for long-term degradation forecast and analysis. When it comes to long-term degradation forecast, the available models (even the statistical models) still can provide nonrealistic degradation forecasts. This is due to many factors, such as assuming constant degradation factors throughout the module's lifetime, not considering the effects of different degradation factors and putting fewer attention on data treatment and evaluations. These effects were discussed in [56], and a model was proposed taking the above mentioned effects into consideration in order to improve long-term degradation forecast.

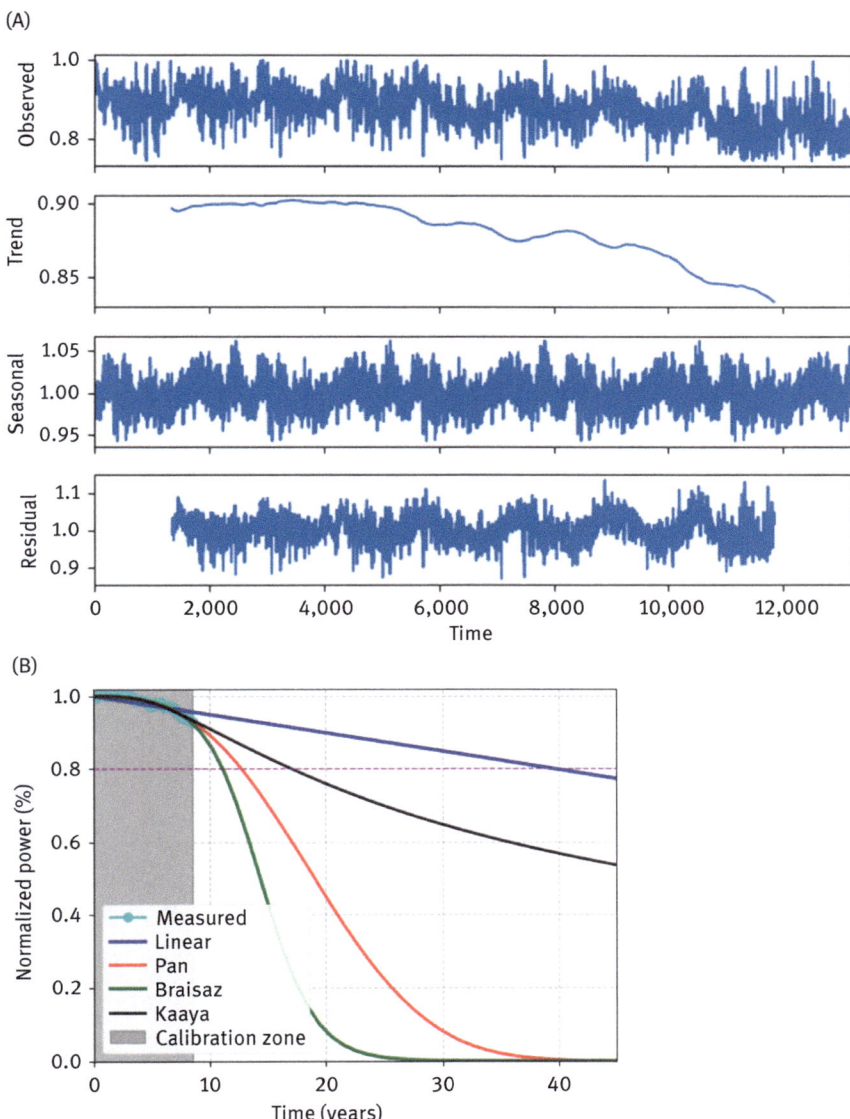

Fig. 9.9: (A) Multiplicative decomposition of a PV systems data showing the four components: observed (measured data), trend, seasonal, and residual. (B) Long-term degradation forecast of the four different reliability models: linear (blue), Pan (red), Braisaz (green) and Kaaya (black). The measured system data that are used to calibrate the models are shown in cyan. The dashed line at 0.8 shows a 20% performance loss of the initial value. Normalization was done on the initial power of the trend.

Part II: **Crystalline silicon module sustainability**

Sina Herceg, Sebastián Pinto Bautista, Karl-Anders Weiß

10 Market-related topics of sustainability

Sustainability describes the principle of a responsible resource use in accordance with society, economy, and the environment (Fig. 10.1). In a product context, all three areas should be addressed equally. In a broader context, the term sustainability is often only used to describe the economic and ecological measures or to focus on the ecological sustainability. In this chapter, the focus lies on the ecological sustainability of photovoltaic (PV) modules.

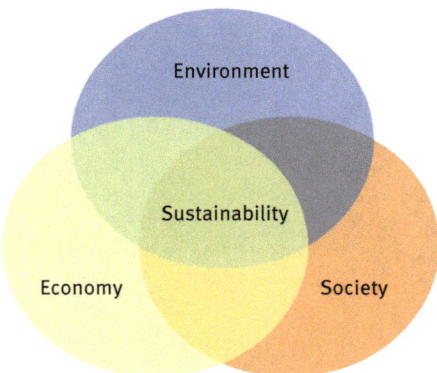

Fig. 10.1: Sustainability including impacts in three different areas.

The development of a circular economy framework, which would allow close loops within the lifecycle of a product through a "cradle-to-cradle" evaluation while providing value creation opportunities, should be based on three waste reduction concepts: reduce, reuse, and recycle (3R's principle). This would ease the accomplishment of better performance with lower cost and environmental impact, in other words, to meet the PV sustainability criteria. In the case of PV, as shown in Fig. 10.2, each stage in the lifecycle consumes resources and energy while releasing emissions and waste to the environment. Reduction of raw material consumption in the early stages of the production line will lower the burden in the following stages.

   The awareness of sustainable production and consumption has grown in recent years due to several developments in policy and society. On the global scale, the 17 sustainable development goals (SDGs) (Fig. 10.3) of the United Nations (UN) followed the millennium development goals' (MDGs) structure and described the measures that have to be taken to ensure sustainable development worldwide on an economic, social, and ecological level.

https://doi.org/10.1515/9783110685558-010

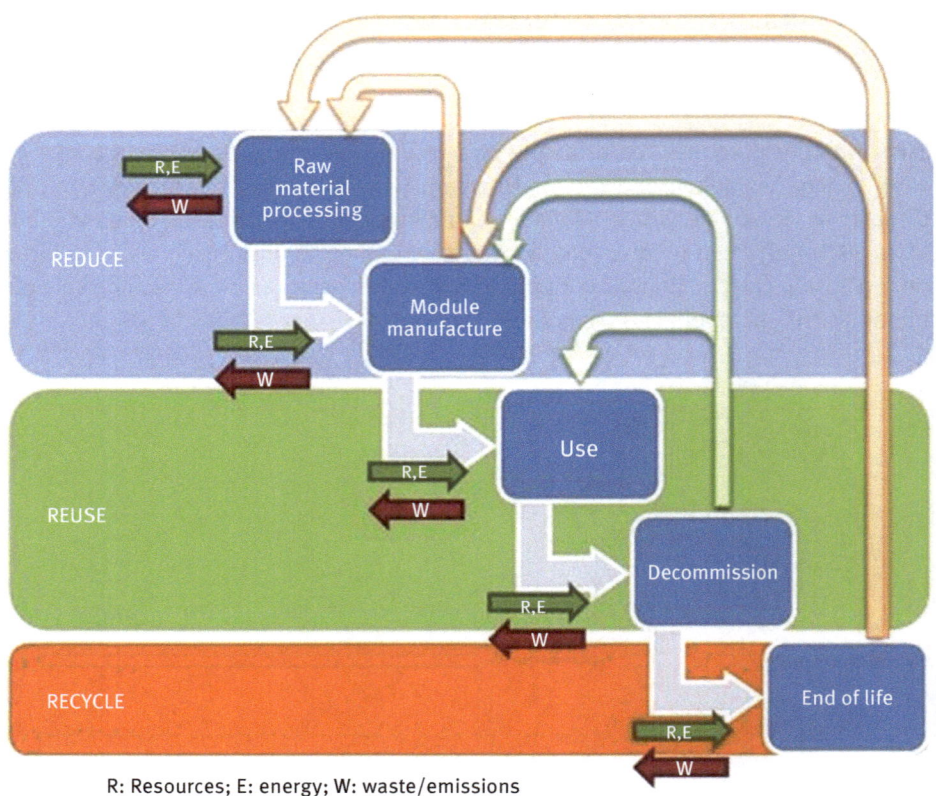

R: Resources; E: energy; W: waste/emissions

Fig. 10.2: Life cycle of photovoltaic modules [75].

Besides contributing to SDG 7 "Affordable and Clean Energy" and SDG 13 "Climate Action," PV industry is more and more aiming to contribute to SDG 12 "responsible consumption and production." Also strategies to reuse materials and to avoid substances with negative impact to the environment address these goals. Some measures to increase sustainability of products already have been transferred in legislation processes that ban products with strong negative impact or that support sustainable developments. It is expected that such legal measures will become prevalent internationally and for all technologies in the near future.

Besides these legal and social developments, market requirements also include sustainability of products, especially products that are linked to sustainability. For example, PV is one of the most prominent sources of renewable (sustainable) energy and is expected to be on the highest level of sustainability. There is also a strong growth of sustainable "green" investments, reaching a multibillion market including

Fig. 10.3: Sustainable development goals of the United Nations.

stock-traded green investment funds, and even some companies or banks already limit their activity to sustainable investments.

The book tries to give an overview on most common sustainability assessment methods for PV (Chapter 11) and influencing parameters (Chapters 12–15) as well as certification and legislation (Chapter 16).

Sina Herceg, Sebastián Pinto Bautista, Karl-Anders Weiß
11 Sustainability assessment methods

Despite photovoltaic (PV) deployment being motivated by an energy transition toward a cleaner grid, the technology is not exempt of producing negative impacts to the environment during production and disposal. The ecological sustainability of a specific system or a product can be measured by looking at its environmental profile. This profile can also be understood as the ecological footprint and serves to describe the interactions and associated impacts on the environment. When evaluating the whole lifecycle of PV modules, hot spots of energy use and materials' demand can be identified. These emissions have to be properly assessed in order to lower the environmental downsides of PV while turning it into an exemplary technology.

11.1 Life cycle assessment (LCA)

Life cycle assessment (LCA) assesses potential environmental impacts of a product, from raw material extraction (cradle) to the production factory (gate), use phase, and final stage – disposal (cradle to grave) or recycling (cradle to cradle). LCA is standardized according to the DIN EN ISO 14040 and 14044 [75, 76] framework and consists of four stages which should be followed in an iterative manner (Fig. 11.1):

1. Goal and scope definition
The first part should define the intended application and the targeted audience. As for scope, several parameters such as the product system, the system boundaries (Fig. 11.2), the functional unit (FU), and reference flows have to be determined. The FU is a reference measure that describes the purpose of the system and that must be defined in accordance with the goal and scope of the assessment. For most practical purposes, an FU of 1 kWh of electricity is normally employed when PV systems or modules are evaluated and thus the environmental profile will be defined as a function of this unit. Depending on the defined boundaries, the system can be limited to photovoltaic modules, or might as well comprise Balance-of-System (BOS) components such as inverters, cabling and mounting structures required to install a fully functional PV plant.

2. Inventory analysis
Based on the initially defined goal and scope, the inventory analysis includes data collection and the quantification of relevant inputs and outputs to the product system according to the functional unit. According to the iterative of approach of an LCA, goal and scope definition might have to be modified accordingly when new requirements or limitations are identified during the data collection.

https://doi.org/10.1515/9783110685558-011

3. Life cycle impact assessment (LCIA)
LCIA is the final quantitative phase of LCA, where the inventory results are assigned to impact categories (classification) and translated into impacts (characterization). According to ISO 14044, the LCIA includes both mandatory and optional elements. Mandatory elements are the selection of impact categories, category indicators and characterization models, classification (assignment of LCI results to impact categories), and characterization (calculation of category indicator results). Whereas the optional elements of normalization, grouping, and weighting can be helpful for easier interpretation and communication of results, they also add an additional level of uncertainty.

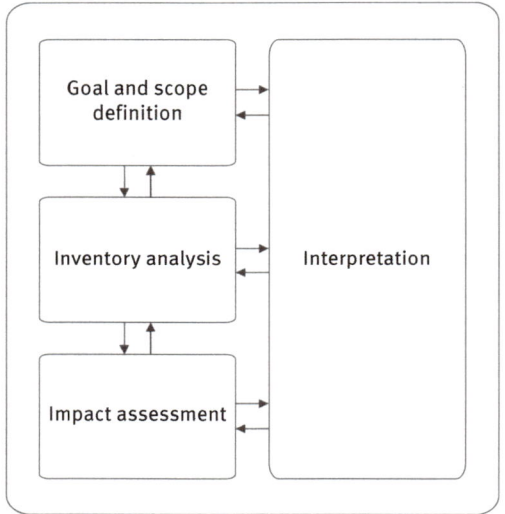

Fig. 11.1: LCA framework according to ISO 14040.

4. Interpretation
In the interpretation phase, environmental hot spots will be first identified and then discussed according to the results from the inventory analysis and impact assessment phases. In the next step, the study's completeness and consistency is evaluated. Based on the identification of the most significant impacts and the evaluation of the results, conclusions can be drawn. Moreover, the limitations of the study can be identified and thus recommendations can be provided for the improvement of both the study and the environmental performance of the product.

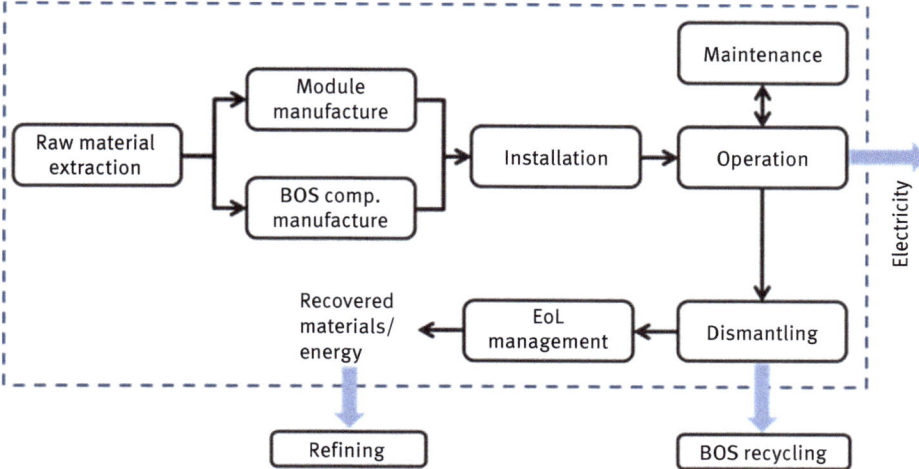

Fig. 11.2: General system boundaries describing the whole life cycle of a PV plant.

11.2 Impact assessment methods

LCIA is the phase of LCA where the inventory results are assigned to impact categories (classification) and translated into impacts (characterization). The impact assessment method should be chosen depending on the goal and scope of an LCA
study. For example, the method of cumulated energy demand (CED) is widely used
for energy producing technologies to further assess energy payback times (EPBT) or
energy returned on energy invested (EROI). For the specific case of PV electricity,
the assessment can be performed in accordance with the European Product Environmental Footprint (PEF) recommendation from the European commission, which
establishes a set of 15 impact categories related to the International Reference Life
Cycle Data System methodology [77] described in Tab. 11.1. The relevance of each
impact category depends on the goal of the study and the context around it but it is,
however, often subject to subjective value judgments.

Tab. 11.1: Environmental impact categories evaluated in the LCA of PV systems [78].

Indicators from the PEF guide		
Climate change	Radiative forcing as global warming potential	kg CO^2 eq·
Ozone depletion	Ozone depletion potential	kg CFC-11 eq
Human toxicity, cancer effects	Comparative toxic units for humans	CTUh, c
Human toxicity, noncancer effects	Comparative toxic units for humans	CTUh, n-c
Particulate matter/respiratory effects	Intake fraction for fine particles	kg PM2.5 eq
Ionizing radiation, human health	Impact of human exposure relative to U^{235}	kBq U^{235} eq
Photochemical ozone formation	Tropospheric ozone concentration increase	kg NMVOC eq
Acidification	Accumulated exceedance	mol H^+ eq
Eutrophication, terrestrial	Accumulated exceedance	mol N eq
Eutrophication, freshwater	Fraction of nutrients reaching freshwater end compartment	kg P eq
Eutrophication, marine	Fraction of nutrients reaching marine end compartment	kg N eq
Freshwater ecotoxicity	Comparative toxic unit for ecosystems	CTUe
Land use	Soil organic matter	kg C deficit
Resource depletion, water	Water abstraction related to the local scarcity of water	m^3 water eq
Resource depletion, mineral, fossil, and renewable	Scarcity	kg Sb eq

Figure 11.3 shows the sources of environmental burdens of a specific PV system separated between modules and Balance of System (BOS) components.

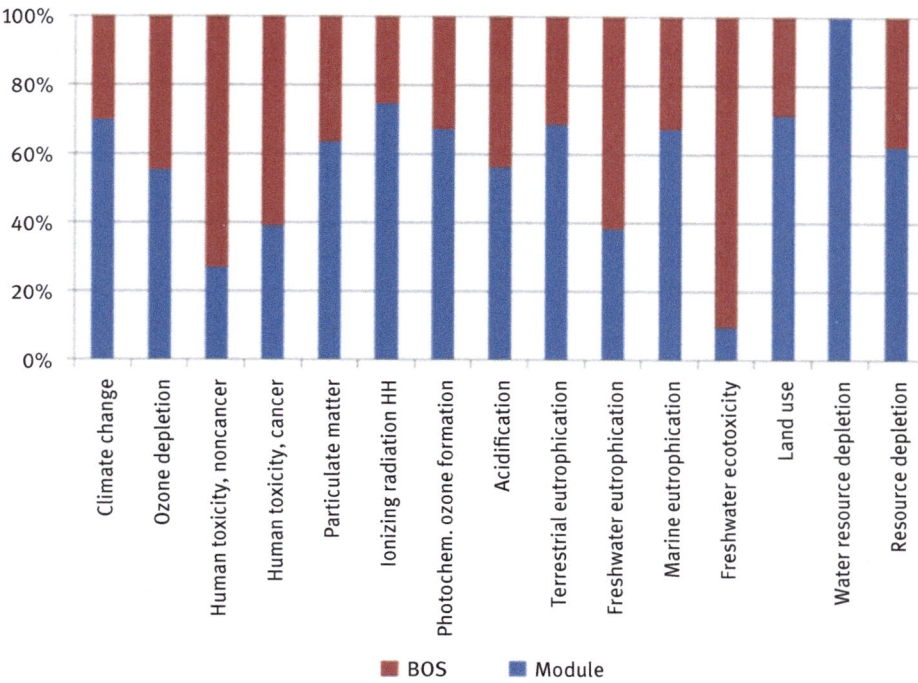

Fig. 11.3: Burden distribution of the ecological footprint of a PV system.

11.2.1 Embodied energy and energy payback time (EPBT)

One commonly known assessment method is CED, which is widely used for a range of products because energy consumption is often strongly correlated with the carbon footprint of products:

$$\mathrm{CED} = \mathrm{CED_P} + \mathrm{CED_U} + \mathrm{CED_D} \tag{11.1}$$

where CED represents the total cumulated energy demand, $\mathrm{CED_P}$ is the cumulated energy demand for production, $\mathrm{CED_U}$ is the cumulated energy demand while usage, and $\mathrm{CED_D}$ is the cumulated energy demand for disposal.

CED can then be further used to calculate EPBT. The EPBT is the time the system needs to generate an amount of energy that equals its CED. Neither the CEA nor the EPBT includes the systems' actual lifetime and therefore gives no information if a system can produce energy way beyond its EPBT or will be at its end of life shortly after this break-even point:

$$\mathrm{EPBT_{PV}} = \mathrm{CED_{PV}}/E_{\mathrm{PV}} \tag{11.2}$$

where $EPBT_{PV}$ is the EPBT of the PV module, CED_{PV} is the total cumulated energy demand of the PV module, and E_{PV} is the annual energy yield of the PV module.

Third, the EROI can be used to express the relation between lifetime energy output and CED:

$$EROI_{PV} = E_{PV\ Total}/CED_{PV} \qquad (11.3)$$

where $EROI_{PV}$ is the energy return on invested, $E_{PV\ Total}$ is the lifetime energy yield of a PV module, and CED_{PV} is the total cumulated energy consumption of a PV module.

11.2.2 Carbon footprint and other impact categories

There are a number of different models for LCIA. Basically, they can be divided into two categories: midpoint analyses, which describe direct effects on the environment and their causes (e.g., acidification or global warming), and endpoint analyses, which add more complexity (but also uncertainty) by using further normalization, grouping, and weighting steps to describe the total resulting impact on specific areas (e.g., on human health or ecosystems).

The European Union's Product Environmental Footprint Category Rules (PEFCR) [77] identify the following impact categories as "most relevant" for silicon PV modules because they cumulatively contribute at least 80% to the total normalized and weighted impacts:

– Particulate matter
– Climate change
– Resource use, fossil fuels
– Resource use, minerals, and metals

As most relevant life cycle stages, the following have been identified in the PEFCR because they cumulatively contribute at least 80% to the characterized results in each of the most relevant impact categories:

– Raw material acquisition and preprocessing
– End of life

Sina Herceg, Sebastián Pinto Bautista, Karl-Anders Weiß
12 Life cycle impacts and embodied energy

The lifetime energy demand of photovoltaic (PV) technology can be split up into energy used during production and resource extraction, during operation and maintenance, and during end-of-life recycling and disposal. In the case of PV, life cycle impacts mainly occur during production, but also the end-of-life stage should be assessed carefully.

12.1 Resource extraction and production

The supply chain of materials used for module manufacturing is responsible for the release of several emissions to the environment. Lead used for soldering and in metallization pastes as well as fluorinated backsheets are not only a problem during production, but regarding end-of-life treatment of the modules. The production also requires auxiliary substances like dopants, cleaning acids, and etching agents, which can be hazardous to the environment. When it comes to embodied energy, for silicon PV-framed glass-backsheet modules, solar-grade silicon is the most energy-intensive material to be processed (Fig. 12.1). Cell production is responsible for around 70–80% of the modules' cumulated energy demand (CED). Another 7–8% come from the aluminum frame as well as up to 5% from the solar glass. The rest of the embodied energy is allotted to polymeric fractions like backsheet and encapsulation as well as to module assembly and other metals.

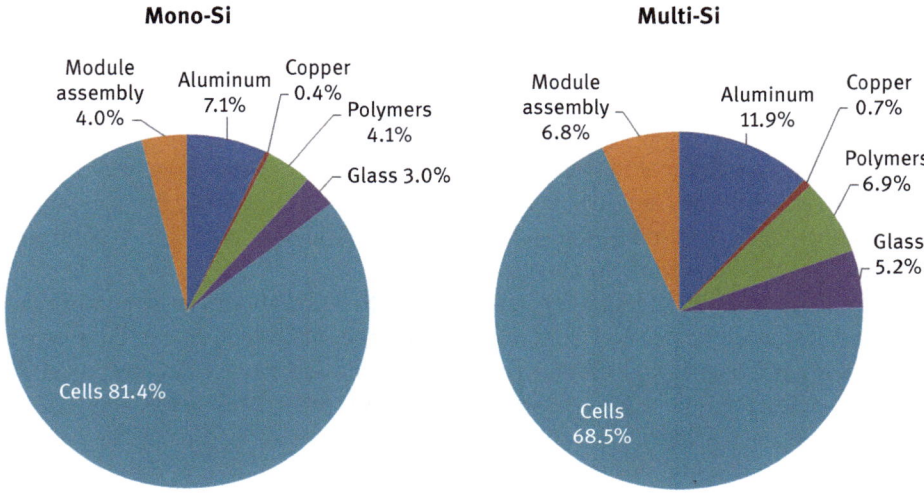

Fig. 12.1: Embodied energy of PV modules [77].

https://doi.org/10.1515/9783110685558-012

Looking at the PV system, the modules contribute to 70–80% of the overall CED, whereas the mounting system equals to around 9–14%, the inverter is between 7% and 10%, and the electric installation makes up to 2–4% of the CED (Fig. 12.2).

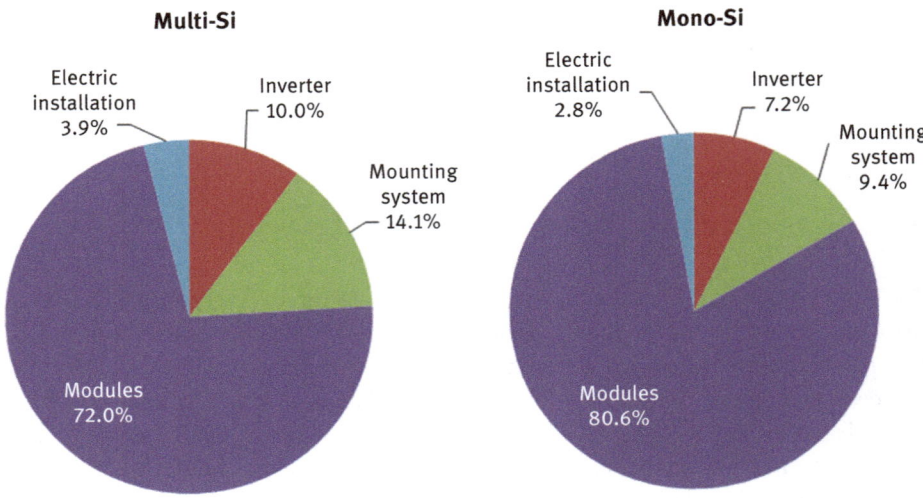

Fig. 12.2: Embodied energy of PV systems [77].

Most efforts to further reduce the environmental impact of PV are located in this life cycle stage. Although originally often driven by economic reasons, measures such as process optimization, efficiency increase, or material reduction have a great potential to also optimize PV modules in an environmentally friendly manner.

12.2 Operation and maintenance

During operation and maintenance, energy output exceeds the energy used by auxiliary energy, cleaning, or minor repairs by far. This is in contrast to fossil technologies like gas- or coal-fired power generation where the operation phase is usually the one with the highest environmental impacts.

This part of the PV lifecycle is often treated with less elaborateness than production and disposal phase, regardless of the effects that maintenance can have on aspects like degradation and lifetime and thus on the lifetime energy output of PV technologies. All the measures described in Part I of the book aim at increasing service life and decreasing degradation of PV modules and so also PV systems. Both approaches lead to higher lifetime yield. Since the yield in kWh is the reasonable functional unit of PV when sustainability is assessed, improved reliability directly influences sustainability. Yield is in the denominator of the relevant equation, so higher lifetime yield improves sustainability – of course depending on the taken measures and the related environmental impacts.

Inspection, replacement of wear parts, or cleaning can improve energy output as well as the systems lifetime, whereas their impacts seem to be comparably small. Some maintenance measures have direct and significant impact on the yield. Especially cleaning measures of plants in areas with high soiling loads have to be mentioned in this context. While the positive effect on yield can be determined relatively easy in this case, the environmental burdens of cleaning are not as easy to determine. Depending on the cleaning technology, water is necessary, often even demineralized water which is usually not available at sites with high soiling loads, which are typically in arid regions. Desalination and potentially also transportation of water can cause significant environmental impacts.

Unfortunately, no reliable life cycle inventory data or life cycle assessment studies exist to the author's knowledge to quantify the actual impacts of maintenance of PV modules.

Sina Herceg, Sebastián Pinto Bautista, Karl-Anders Weiß

13 End-of-life treatment

Circularity of resources requires close loops within the lifecycle of a product through a "cradle-to-cradle" approach for – if possible all – materials. With focus on waste reduction this means in practice reduce, reuse, and recycle (3R's principle) and the accomplishment of better performance with lower cost and environmental impact, in other words, to meet PV sustainability criteria. In the case of photovoltaics (PV), as shown in Fig. 13.1, each phase of the lifecycle consumes resources and energy while releasing emissions and wastes to the environment. So reduction of raw material consumption in the early stages should be addressed as well as reuse and recycling in the end-of-life (EoL) phase.

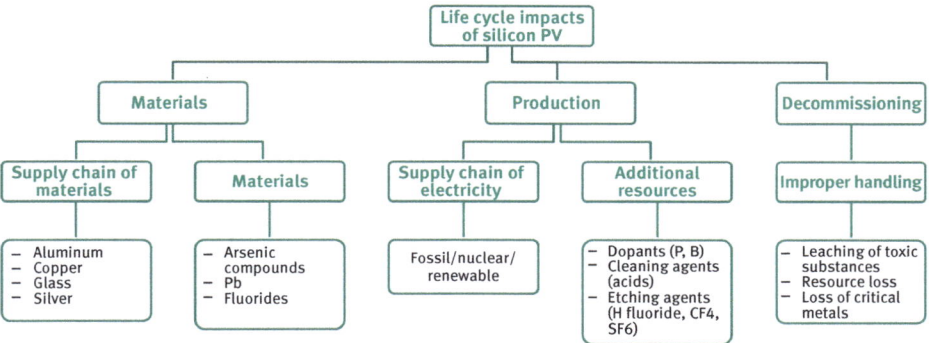

Fig. 13.1: Life cycle of photovoltaic modules [79].

To reach the EoL stage, a module may experience regular performance degradation during the expected lifetime until reaching e.g. 80% of the initial performance, as defined by most manufacturers. Specific lifetimes may vary between types of technologies; however, the most used values in literature and in the market are 25–30 years. The complexity of the degradation mechanisms makes it difficult to predict accurate life spans having as a result modules that still perform over this threshold even after fulfilling their life expectancy.

Handling of products reaching their EoL can be done in several ways: comprising reuse in further applications, refurbishment of some components, recycling of materials, incineration, and landfilling. Proper handling of the material is in any case a vital task when assessing and improving the environmental performance of any product. This section describes the main aspects involved when managing PV waste with a special remark on recycling.

https://doi.org/10.1515/9783110685558-013

13.1 Recycling – material recovery

It is to be expected that the global PV panel waste will scale up to 60–80 billion tons of material by 2050 [80]. However, until now only small flows of PV waste had to be handled. This has restrained the consolidation of a specialized waste processing industry due to a lack of profitability. PV waste management will gain relevance proportionally to the amounts of waste that are expected to arise with the phasing out of old installations in the upcoming years and decades.

Currently, PV panels are often treated in facilities specialized in glass or electronic waste recycling through mechanical processing such as shredding and crushing of the modules. More dedicated recycling approaches that consider the recovery of materials like silicon and silver or the proper treatment of backsheets have the potential to significantly lower the PV modules' environmental footprint [81]. By conserving and reusing raw materials like metals and semiconductors, the resource use for new module manufacturing can be reduced, whereas the environmental impacts of the recycling process of c-Si PV modules are only around 1% compared to the impacts caused by the production of a roof-mounted plant [80].

In order to include the EoL phase, an evaluation of the effects of implementing a waste management scheme on the ecological footprint of PV electricity production has been performed. Two different schemes have been modeled, corresponding to what here has been named "Basic recycling" and "Dedicated recycling." The first one refers to the already implemented waste management approaches limited to recover only bulk materials – such as aluminum and glass – by simple processes in order to comply with the European legislation of mass recovery. These approaches leave behind other valuable materials due to the added complexity of such activities. Currently, this type of approach is performed mostly in laminated glass, metal, and electronic waste recycling plants where PV modules are also suitable to be processed. In addition to basic bulk recovery, some research programs have intended to develop dedicated processes that allow recovery of other valuable materials present in the modules that cannot be retrieved by conventional means. Such is the case of silicon wafers, silver, and other embedded materials, which even though being present in very small amounts, given their scarcity and market price entail high potential value.

The recycled materials are later refined and reintroduced into different markets, which translates into avoiding burdens from raw material extraction. This is accounted for in the form of environmental credits which, when added up with the burdens from recycling itself and with the impacts from electricity production, modify the overall ecological footprint, normally reducing it since they have opposite signs (+/–). The recycling approaches presented here concern only the processing of waste of PV modules and leave out potential recycling of BOS components, which would result in additional benefits.

From Fig. 13.2 it becomes clear that in general recycling leads to improving the ecological footprint, but these burden reductions are relevant mostly when dedicated

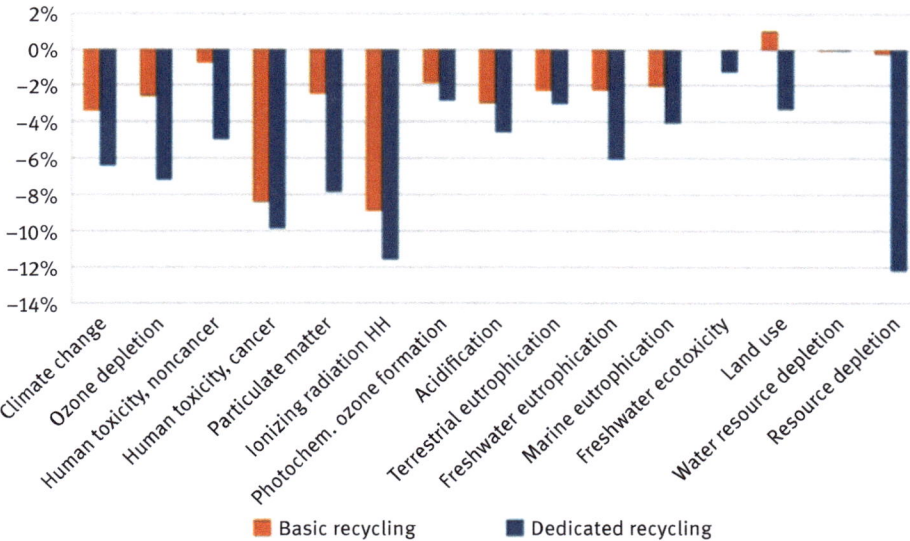

Fig. 13.2: Influence of waste management schemes on the footprint of a PV system.

recovery is performed. The results adopt, in most cases, the form of negative values, which represent the percentage points that the overall footprint of PV electricity could be reduced from the reference (baseline) scenario. In the impact category "Land use," the positive value means that basic recycling leads to actual burdens that are to be added up to the initial profile from electricity production.

The potential carried by dedicated recycling approaches is higher in every category. Specifically, a great amount of benefits are caused due to the avoided need of primary silver that is instead being recovered from the modules. This is particularly beneficial in the "Resource depletion" category. In addition, the recovered silicon can be reintroduced in the PV production chain as metallurgical-grade silicon, leading to avoided consumption of energy which makes these approaches highly appealing from the ecological perspective. However, a profitable mass scaling of these techniques greatly relies on the availability of big flows of waste to be treated.

13.2 Energy recovery

Although energy can already be recovered from the gas emissions produced by cumulated organic matter in landfills, specialized thermal processes such as incineration and pyrolysis, characterized by the presence or absence of oxygen, respectively, seek to reduce waste in terms of weight and volume while releasing energy in the form of heat or storable fuel. These processes aim to recover the embedded energy within a material, avoiding the burdens of energy production by any other conventional means.

13.3 Landfilling

Under the waste management framework, landfilling refers to the disposal of waste that cannot be processed by any other means such as recycling, composting, and incineration, due to inherent properties of the material or because of deficiencies in the local waste management scheme. A modern landfill is a carefully engineered deposit where waste is isolated from the rest of the environment. This isolation prevents leachate (mix of rainwater and water soluble compounds present in the waste) to penetrate into the soil or groundwater reservoirs that could generate potential risks to the environment and human health. Landfills are classified according to the type of waste that is being deposited: (a) municipal solid waste, composed mainly by household wastes and other nonhazardous wastes; (b) industrial wastes such as demolition materials or coal combustion residuals, and (c) hazardous wastes which contain specific toxic substances mainly from industrial processes (EPA 2018). For PV modules, only small fractions like residues of glass, metals and foils as well as ashes from energetic recovery will be sent to landfill.

Sina Herceg, Sebastián Pinto Bautista, Karl-Anders Weiß

14 Approaches to improve sustainability

In order to improve the environmental profile of photovoltaics (PV), different strategies could be addressed, from reducing energy consumption during manufacturing
by means of an optimized production line to reducing transport distances through
an optimized distribution network. However, high sensitivity of the ecological footprint of the system to degradation rates is expected. Strategies aiming to extend the
service life of the modules could bring significant environmental benefits.

By means of a life cycle assessment, it becomes also possible to assess the potential that lies within reliability and service life optimization. Figure 14.1 displays
the potential reduction of the ecological footprint of the produced electricity of a PV
system for the main impact categories when assuming different technical optimization scenarios. The results are presented as a percentage of the baseline, that is, a
PV plant with multi-Si modules with a lifetime of 30 years and a linear degradation
rate of 0.7% per year. An ideal module that does not suffer from any degradation
and nevertheless has a fixed lifetime of 30 years would lead to a 10% reduction of
the electricity's footprint in every category. A different scenario, assuming that the
modules are allowed to operate below the benchmark of 80% of the initial performance reaching a 40 years' lifespan, already lowers the footprint per kWh down to

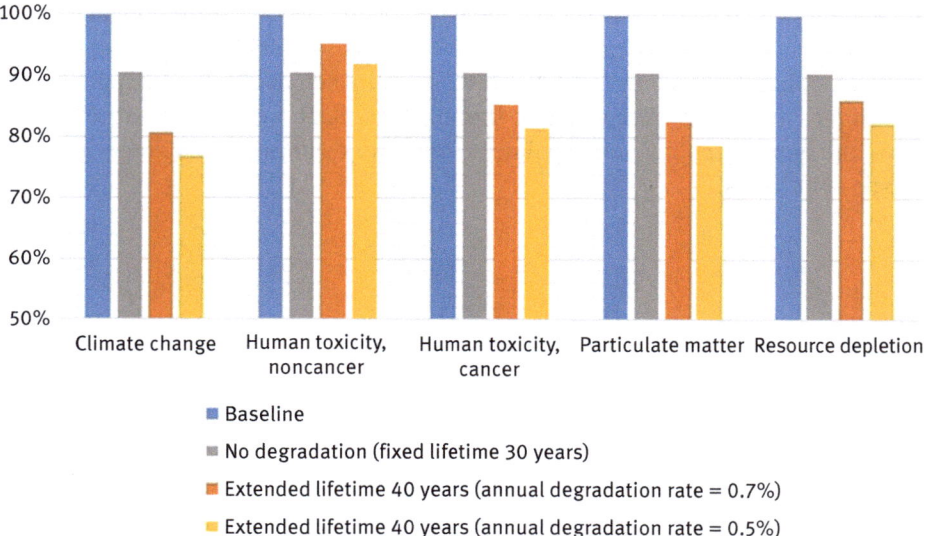

Fig. 14.1: Potential for reduction of ecological footprint of PV electricity produced by a hypothetic
system. Four scenarios representing improved reliability are calculated and the effect on five major
impact categories is shown.

https://doi.org/10.1515/9783110685558-014

about 80–95% with respect to the baseline. One last scenario displays the potential benefits of lowering the annual degradation rate to 0.5%, thus allowing the modules to produce electricity over 40 years while operating above the 80% suggested performance limit. The latter could potentially bring the ecological footprint of PV electricity down to about 76–92% with respect to the baseline. For all these scenarios, the lifetime of the Balance-of-System (BOS) components remains constant, hence, proportionally increasing their demand. It has to be mentioned that the scenarios are hypothetical and are neither related to a specific technological improvement nor a specific module type.

Each measure to improve sustainability should be analyzed regarding the related environmental impacts as well as regarding the effects on yield before they are realized. Not all measures that seem to be obviously beneficiary for sustainability have only positive effects; sometimes, the effects are negligible or even negative. This can especially be the case when materials are replaced by substitutes with lower environmental impact. If this leads to reduced reliability and lifetime or higher degradation, the effect can in the end be opposite.

There is also plenty of room for technical improvement of the PV waste management. Even when the initial estimations seem to entail moderate contributions to the ecological footprint of electricity, further development of these and new techniques will turn recycling into a relevant tool to make PV systems an exemplary technology for the energy transition. Additionally, recycling should still be studied as a whole, including the processing of the BOS components as well as the environmental credits that come with it. This will most likely lead to a greater reduction of the electricity production's footprint. The development of an efficient recycling network will not only bring environmental gains but will also have a great impact on the economics and financial aspects of the logistics. Such an assessment must also take into account policy and legislative framework so that the outcomes meet realistic conditions.

Sina Herceg, Sebastián Pinto Bautista, Karl-Anders Weiß

15 Sustainability certification and legislation

Rapidly growing markets require the development of standards, which are able not only to bring industries within the international market to common terms but also to include environmental aspects that ensure safety and reduced hazards during the different stages of the photovoltaic (PV) life cycle. They allow cost-effectiveness to producers willing to participate in several national markets, avoiding the need to meet a different set of requirements for each region that would otherwise end up in increased production costs. Standards also serve as a guide during the development state of new technologies or materials, indicating which tests and parameters the product will have to comply with. The standardization process has to cover all the different stages of PV module production, starting from the raw materials processing and supply to the manufacturing of the cells and assembling of the modules. The main topics addressed by standards relate to quality, reliability, safety, sizing, characterization procedures, and material/component specifications among others. Some of the most relevant sustainability related standards that have been developed to attend the different needs of the PV production chain and that are more commonly required by the market are discussed in detail further. A description of standards addressing reliability and quality of PV modules can be found in Chapter 8.

Some regulatory frameworks have been established in order to improve the environmental performance of PV technologies, either by changing design concepts or to promote a better management of the waste material.

15.1 European regulations and measures

There are several legislation and certification measures that have been introduced by the European Union (EU) to support sustainability of PV or measures that are discussed at the moment.

15.1.1 Waste Electrical and Electronic Equipment (WEEE) directive

A specific framework for the waste of end-of-life (EoL) material has been implemented in the EU. The Waste Electrical and Electronic Equipment (WEEE) directive is in charge of regulating waste management, setting minimum requirements regarding aspects such as collection, treatment, and financing of the processes required while seeking to reduce the volumes of electronic waste produced. Despite entering into force in early 2003 it was just until 2012 when the directive was revised to include specific aspects of

https://doi.org/10.1515/9783110685558-015

PV panel waste management, which were meant to be transposed into national law by early 2014. Each country within the EU became responsible for implementing a waste management scheme in accordance with the directive with freedom to include specific and stricter policies. To manage the waste stream, the WEEE promotes the extended producer responsibility principle, which obliges PV producers (manufacturers, distributors, resellers, importers, and distance sellers) to take care of the collection and disposal of their products when they reach EoL as well as for financing the implemented model to do so. More specifically, producers are liable of the following responsibilities [81]:

Financing responsibility: producers must become a part of a collective or individual scheme able to offer financial guaranty for the establishment of collection points and specialized treatment facilities for recycling.

Reporting responsibility: producers have to report periodically the number of sold, collected, and treated panels.

Information responsibility: producers must properly label their products in compliance with the directive, informing customers how and where to do the disposal of the panels as well as including handling guidelines for collection, storage, dismantling, and treatment including information of hazardous material content and potential risks.

The directive has also established particular targets for collected, recovered, and recycled mass of waste as specified in Tab. 15.1, while taking into consideration specific material-dependent amounts. Recovery refers to physical operations leading to the availability of material streams while recycling makes reference to the final usable material.

Tab. 15.1: WEEE targets for disposal of PV waste.

Period	Collection	Recovery	Recycling
2016–2018	45%	80%	70%
From 2018 on	65%	85%	80%

To determine the financing responsibility, one must classify the end customer as private or nonprivate household. When the customer is a private household (or has a similar demand), the transactions with the producer are called business-to-consumer (B2C) and the producer is not allowed to enter into contractual arrangement, whereas for others these transactions receive the name of business to business (B2B). This framework allows each state covered by the WEEE directive to designate the financial liability of the stakeholders. The first approach of financing contemplates individual funding and collective liability (also named joint and several) schemes for B2C transactions, and the responsibility lies on the producer. Individual ones when being prefunded have not shown cost-effectiveness and therefore are normally addressed through the execution of a pay-as-you-go model where the waste management costs are covered when waste occurs and being supported by the "last man standing" insurance, which serves as a backup to cover the costs in case all the producers disappear from the market. The second approach allows contractual arrangements between producer and customer under the context of B2B transactions, since both of them may be in the condition of covering the EoL waste management demands. This is the case for large PV plants which are best positioned to perform effective collection and recycling to comply with the WEEE requirements.

In Germany, as an example for a mature PV market with specific regulations, the WEEE directive was included in the national law through the Electrical and Electronic Equipment Act, also known as ElektroG in late 2015. Waste management is regulated by the National Register for Waste Electrical Equipment, which is in charge of registering waste producers, ensuring that they participate in the internal setting of rules, reporting to the Federal Environment Agency annual flows of material and waste amount, and coordinating provision of containers for the waste takeback. The remaining operational tasks, from collection to recycling and disposal, are in charge of the producers. In addition, for B2C transactions, the collective liability scheme defines different levels of operation and financing: the first level includes the operation of a collection and recycling system as well as covering the related costs for the processing of products that were placed in the market before the implementation of regulations. The second level contemplates future financing related to the waste management costs of products placed in the market after implementation of the law. Waste amounts to be processed are designated to each producer depending on the volumes collected and on their current market share. Additionally, if a producer disappears from the market, its responsibilities will be assigned to the remaining ones.

15.1.2 Ecodesign, Ecolabel, energy label, and Green Public Procurement

In the EU, the "Ecodesign directive" was first implemented in 2009, setting mandatory requirements that energy-related products have to meet and consider within the initial stages of product development if they intend to have a place in the market. In such a way, the least performing products are eliminated from the market leading to reinforced industrial competitiveness and innovation, contributing at the same time to achieve the EU's energy efficiency and environment protection goals (Fig. 15.1). Among the product aspects covered under this framework, features such as durability, reparability, design for disassembly, labeling, ease of reuse/recycling, and general emissions can be found, and some of which greatly depend on the initial design of the product. The evaluation of the products follows the Methodology for Ecodesign of Energy-Related Products (MEErP), based on the conduction of different life cycle assessments for the identification of hot spots in the life cycle of a product [82].

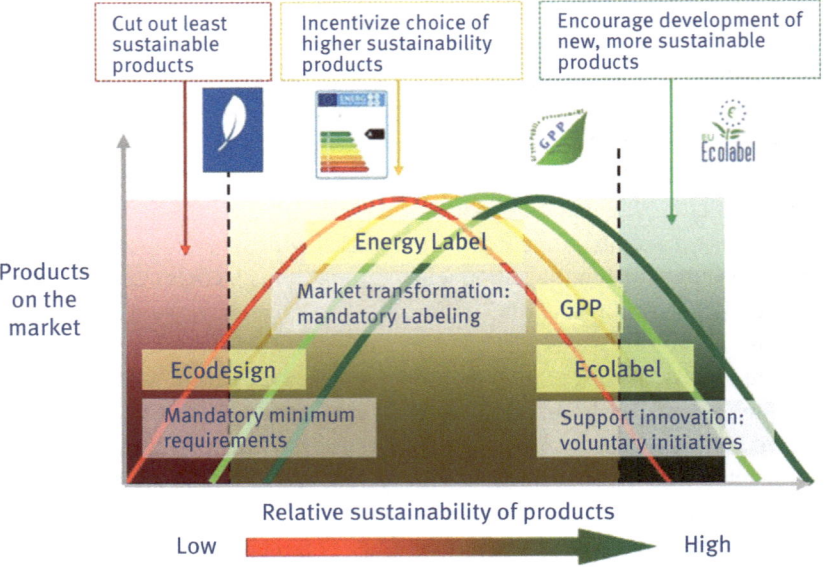

Fig. 15.1: Overview on EU product policy instruments, targets, and sustainability levels. Source: EC JRC Seville Technical Report, Seville, 2018, JRC114333.

In addition, the Ecolabel scheme complements the goals of this directive. Comprising a voluntary approach, it seeks to promote products with increased performance and low impact through their whole life cycle, standing out among the others and being rewarded by the market dynamics. The criteria used for this labeling are based on technical evidence, including usage of hazardous substances, resource efficiency, durability, and recyclability.

Another rating system to display the sustainability of products is the EU energy label, which has been mainly developed to rate products according to the amount of energy they use. Since PV systems do not consume energy during operation but generate electrical energy, a direct application seems not to be obvious but the rating systems could be adapted in a way that systems with high yield are highly ranked.

The fourth system to support sustainable products in the EU is called Green Public Procurement and addresses – as the name indicates – specifically public procurement processes. It, therefore, describes a set of conditions and requirements products have to fulfill to be listed as "green" to make it easier to include these requirements in procurement processes.

The specific case of PV (including panels and inverters) has been introduced in the Ecodesign working plan of 2016, and a so-called preparatory study including stakeholder involvement has been performed during the last years. Since the process was – and still is – open, which kind of measure will be applied for PV is still (March 2021) not decided. The preparatory study was not linked to one of the described measures but collected information on the PV market and technologies and targeted to identify the best suitable measure(s) to promote sustainability in PV. The study is in its final phase, and discussions with several stakeholder groups are ongoing at the moment. It is expected that recommendations for establishing one or more of the abovementioned measures are published within the coming months.

The authors expect that at least one of the measures will be implemented and that the formal process for the establishment of the measure(s) will start soon. The time until such measures come into force is different since they require different processes. Ecodesign requires a complete legislation process which typically takes years, while Ecolabel can be established by the Ecolabel board and can be realized within months. It is expected that the labels will be accepted by the industry, as indicated in Chapter 10, and that they will raise solar industry's "green credibility."

15.2 EPEAT and the Sustainability Leadership Standard in the United States

An example of an approach comparable to the EU Ecolabel can be found in the United States under the NSF/ANSI 457-2019 Sustainability Leadership Standard for Photovoltaic Modules and Photovoltaic Inverters [83]. This framework aims at establishing product sustainability criteria and performance indicators for companies that demonstrate sustainability leadership in the market and is based on a voluntary approach and is backed up by a third-party verification scheme. It establishes a bronze to gold standard where only a few products are expected to meet the highest level (gold). The standard addresses multiple environmental performance indicators like energy efficiency and greenhouse gas reduction, design for repair, reuse, and recycling, EoL management as well as social responsibility.

The US-based Green Electronics Council is managing a life cycle-based, third-party-validated Ecolabel for IT and electronic products called EPEAT, and has launched the category "Photovoltaic Modules and Inverters (PVMI)" in 2020. In order to be certified, products have to meet a number of required criteria. There are also optional criteria, which can be fulfilled in order to reach the "gold" standard.

15.3 Carbon footprint in French tendering system

Like a lot of countries, France has established a tendering system to regulate the incentives for the development of renewable energy plants. Unlike many other countries, the French tendering systems is not only rewarding the most cost-effective candidate but also considers the carbon footprint of the plant to be built.

References

[1] D. M. Chapin, C. S. Fuller, and G. L. Pearson. A new silicon p-n junction photocell for converting solar radiation into electrical power, *Journal of Applied Physics*, 25 (5), 1954, 676–677.

[2] E. D. Dunlop and D. Halton. The performance of crystalline silicon photovoltaic solar modules after 22 years of continuous outdoor exposure, *Progress in Photovoltaics: Research and Applications*, 14, 2006, 53–64.

[3] H. Laukamp, et al., Reliability issued in PV systems – experience and improvements, 2nd World Solar Electric Buildings Conference, Sydney, 2000.

[4] S. Kaplanis, et al., Energy performance and degradation over 20 years performance of BP c-Si PV modules, *Simulation Modelling Practice and Theory*, 19, 2011, 1201–1211.

[5] M. Vázquez, et al., Photovoltaic module reliability model based on field degradation studies, *Progress in Photovoltaics: Research and Applications*, 16, 2008, 419–433.

[6] C. Dechthummarong, et al., Physical deterioration of encapsulation and electrical insulation properties of PV modules after long-term exposure in Thailand, *Solar Energy Materials and Solar Cells*, 94, 2010, 1437–1440.

[7] IEC 62788-1-6:2017 Measurement procedures for materials used in photovoltaic modules – Part 1–6: Encapsulants – Test methods for determining the degree of cure in Ethylene-Vinyl Acetate.

[8] Ch. Hirschl, L. Neumaier, S. Puchberger, W. Mühleisen, G. Oreski, G.C. Eder, R. Frank, M. Tranitz, M. Schoppa, M. Wendt, N. Bogdanski, A. Plösch, and M. Kraft. Determination of the degree of ethylene vinyl acetate crosslinking via Soxhlet extraction: Gold standard or pitfall? *Solar Energy Materials and Solar Cells*, 143, 2015, 494–502. doi:10.1016/j.solmat.2015.07.043.

[9] C. Hirschl, M. Biebl-Rydlo, M. Debiasio, W. Mühleisen, L. Neumaier, W. Scherf, G. Oreski, G. Eder, B. Chernev, W. Schwab, and M. Kraft. Determining the degree of crosslinking of ethylene vinyl acetate photovoltaic module encapsulants – A comparative study. *Solar Energy Materials and Solar Cells*, 116, 2013, 203–218. doi:10.1016/j.solmat.2013.04.022.

[10] G. Oreski, A. Rauschenbach, C. Hirschl, M. Kraft, G. C. Eder, and G. Pinter. Crosslinking and post-crosslinking of ethylene vinyl acetate in photovoltaic modules. *Journal of Applied Polymer Science*, 134 (23), 2017, art. no. 44912. doi:10.1002/app.44912.

[11] M. Knausz, G. Oreski, G. C. Eder, Y. Voronko, B. Duscher, T. Koch, G. Pinter, and K. A. Berger. Degradation of photovoltaic backsheets: Comparison of the aging induced changes on module and component level. *Journal of Applied Polymer Science*, 132 (24), 2015, art. no. 42093, doi:10.1002/app.42093.

[12] B. Ottersböck, G. Oreski, and G. Pinter. Correlation study of damp heat and pressure cooker testing on backsheets. *Journal of Applied Polymer Science*, 133 (47), 2016, art. no. 44230, doi:10.1002/app.44230.

[13] G. C. Eder, Y. Voronko, G. Oreski, W. Mühleisen, M. Knausz, A. Omazic, A. Rainer, C. Hirschl, and H. Sonnleitner. Error analysis of aged modules with cracked polyamide backsheets. *Solar Energy Materials and Solar Cells*, 203, 2019, art. no. 110194, doi:10.1016/j.solmat.2019.110194.

[14] K. J. Geretschläger, G.M. Wallner, and J. Fischer. Structure and basic properties of photovoltaic module backsheet films. *Solar Energy Materials and Solar Cells*, 144, 2016, 451–456. doi:10.1016/j.solmat.2015.09.060.

[15] G. W. Ehrenstein, G. Riedel, and P. Trawiel. 2004. Thermal analysis of plastics. Theory and practice. Carl Hanser Verlag, Munich. https://www.hanser-elibrary.com/isbn/9783446226739

[16] Energiedispersive Röntgenspektroskopie, Wikimedia, 2013, available at: upload. wikimedia. org/wikipedia/commons/5/5a/Atom_model_for_EDX_DE.svg.

[17] Introduction to Energy Dispersive X-ray Spectrometry (EDS), University of California Riverside, Central Facility for Advanced Microscopy and Microanalysis, 2013.

https://doi.org/10.1515/9783110685558-016

[18] H. J. Bowley, D. L. Gerrard, and I. S. Biggin The use of laser Raman spectroscopy to study the degradation of poly(vinyl chloride), *Polymer Degradation and Stability*, 20 (34), 1988, 257–269.

[19] J. Haunschild. *Lumineszenz-Imaging – Vom Block zum Modul*, Fraunhofer Verlag, Stuttgart, 2012.

[20] ASTM, *1925, 1925–70 – Standard Test Method for Yellowness Index of Plastics*, Pennsylvania, USA, 1925.

[21] C. Peike. *Degradation Analysis of the Encapsulation Material in Photovoltaic Modules by Raman Spectroscopy*, Fraunhofer Verlag, Stuttgart, 2015.

[22] D. C. Harris and M. D. Bertolucci. *Symmetry and Spectroscopy – An Introduction to Vibrational and Electronic Spectroscopy*, Oxford University Press, New York, 1978.

[23] N. S. Nielsen, D. N. Batchelder, and R. Pyrz. Estimation of crystallinity of isotactic polypropylene using Raman spectroscopy, *Polymer*, 43, 2002, 2671–2676.

[24] J. L. Koenig. Infrared and Raman spectroscopy of polymers, *Rapra Review reports*, 12, (1), 2001, 16–26.

[25] D. C. Miller, M. Owen-Bellini, and P. L. Hacke. Use of indentation to study the degradation of photovoltaic backsheets, *Solar Energy Materials and Solar Cells*, 201, 2019, 110082.

[26] E. Herbert, W. C. Oliver and G. M. Pharr Nanoindentation and the dynamic characterization of viscoelastic solids, *Journal of Physics D: Applied Physics,* 41 (7), 074021, 2008, 1–9.

[27] M. L. Oyen. Spherical indentation creep following ramp loading, *Journal of materials research*, 20 (8), 2005, 2094–2100.

[28] D. E. Mansour, C. Barretta, L. Pitta Bauermann, G. Oreski, A. Schueler, D. Philipp, and P. Gebhardt Effect of backsheet properties on PV encapsulant degradation during combined accelerated aging tests, *Sustainability*, 12, 2020, 5208.

[29] R. G. Maev, E. Y. Maeva, and V. M. Levin. 1996. The application of scanning acoustic microscopy to dielectric polymer composite material study. In Proceedings of Conference on Electrical Insulation and Dielectric Phenomena – CEIDP '96 (717–720). https://doi.org/10.1109/CEIDP.1996.564576

[30] L. V. Mesquita, D. E. Mansour, P. Gebhardt, and L. Pitta Bauermann. Scanning acoustic microscopy analysis of the mechanical properties of polymeric components in photovoltaic modules, *Engineering Reports*, 2020, e12222, https://doi.org/10.1002/eng2.12222.

[31] L. V. Mesquita, D. E. Mansour, D. Philipp, and L. Pitta Bauermann. Scanning acoustic microscopy as a non-destructive method for the investigation of PV module components. 35th European PV Solar Energy Conference and Exhibition, EUPVSEC Brussels, 2018.

[32] Y. Voronko, G. Eder, M. Weiss, M. Knausz, G. Oreski, T. Koch, and K. Berger. 2012. Long term performance of PV modules: System optimization through the application of innovative non-destructive characterization methods. In Proceedings of the 27th EU PVSEC 3530–3535. https://doi.org/10.4229/27THEUPVSEC2012-4BV.3.41.

[33] M. Köntges, G. Oreski, U. Jahn, M. Herz, P. Hacke, K.-A. Weiss, and et al. 2017. Assessment of Photovoltaic Module Failures in the Field: IEA PVPS Task 13, Subtask 3 Report IEA-PVPS T13-09:2017. Sankt Ursen, CH. Retrieved from http://www.iea-pvps.org/index.php?id=435

[34] C. Peike, S. Hoffmann, I. Dürr, K. A. Weiss, and M. Köhl. The influence of laminate design on cell degradation, *Energy Procedia*, 38, 2013, 516–522.

[35] M. Koehl. Modelling of the nominal operating cell temperature based on outdoor weathering, *Solar Energy Materials and Solar Cells*, 95 (7), 1638–1646, 2011.

[36] P. Huelsmann, M. Heck, and M. Koehl M. Simulation of water vapor ingress into PV Modules under different climatic conditions, *Journal of Materials*, 2013.

[37] F. C. Krebs. *The Different PV Technologies and How They Degrade, Stability and Degradation of Organic and Polymer Solar Cells*, John Wiley & Sons, 2012.

[38] Fundamental Properties of Solar Cells and Pastes for Silicon Solar Cells, Stephenson and Associates, Inc., 2010.

[39] M. Koentges, V. Jung, and U. Eitner. Requirements on metallization schemes on solar cells with focus on photovoltaic modules, in: Proceedings of 2nd Workshop on Metallization for Crystalline Silicon Solar Cells, Constance, Germany, 2010.

[40] P. Huelsmann, K.-A. Weiss, and M. Koehl. Temperature-dependent water vapour and oxygen permeation through different polymeric materials used in PV Modules, *Progress in Photovoltaics: Research and Applications*, 22 (4), 2012, 415–421.

[41] E. E. Stansbury and R. Buchanan. Fundamentals of electrochemical corrosion, *ASM International*, 2000.

[42] T. Graedel and R. Frakental. *Journal of the Electrochemical Society*, 137 (8), 1990, 2385.

[43] C. Heces, J. Alonso, and F. Corvo. *Review of CENIC*, 18, 1987, 2–3.

[44] K. Slamova, R. Glaser, C. Schill, S. Wiesmeier, and M. Koehl. Mapping atmospheric corrosion in coastal regions: Methods and results, *Journal of Photonics for Energy*, 2, 2012, 1.

[45] G. Meira. Modelling sea-salt transport and deposition in marine atmosphere zone-a tool for corrosion studies, *Corrosion Science*, 50 (9), 2008, 2724–2731.

[46] S. Oesch and M. Faller. Environmental effects on materials: The effect of the air pollutants SO_2, NO_2, NO and O_3 on the corrosion of copper, zinc and aluminum. A short literature survey and results of laboratory exposures, *Corrosion Science*, 39 (9), 1997, 1505–1530.

[47] G. Jorgensen. personal correspondence, 2010.

[48] M. Lechner, K. Gehrke, and E. Nordmeier. *Makromolekulare Chemie*, Springer Verlag, Heidelberg, 2009.

[49] M. Köhl, D. Philipp, and K.-A. Weiß. Rundvergleich von UV-Prüfeinrichtung für Photovoltaik-Module, GUS Jahrestagung, 2011, Stutensee.

[50] M. Heck, M. Köhl, D. Philipp, J. Schlieper, and K.-A. Weiß. Investigation of surface temperatures, in: T. Reichert, T., ed., *Natural and Artificial Ageing of Polymers: 3rd European Weathering Symposium*, CEEES Publication No. 8, 2007, ISBN: 978-3-9810472-3-3.

[51] S. Hoffmann and M. Koehl. Effect of humidity and temperature on the potential-induced degradation, *Progress in Photovoltaics: Research and Applications*, 22 (2), 2012, 173–179.

[52] J. Berghold, O. Frank, H. Hoehne, S. Pingel, B. Richardson, and M. Winkler. Potential induced degradation of solar cells and panels. 25th European Photovoltaic Solar Energy Conference and Exhibition / 5th World Conference on Photovoltaic Energy Conversion, 6–10 September 2010, Valencia, Spain.

[53] J. Bierbaum, D. Philipp, S. Stecklum, I. Dürr and K.-A. Weiß. Investigation on Snail Track Formation, Degradation Mechanisms and Raman Spectroscopic Examination of the Corrosion Products, EU PVSec, 2015.

[54] Review of Failures of Photovoltaic Modules – Report IEA-PVPS T13–01:2014, Available at: http://www.isfh.de/institut_solarforschung/files/iea_t13_review_of_failures_of_pv_mod ules_final.pdf

[55] M. Köhl, B. Carlson, G. J. Jorgensen, and A. W. Czanderna. *Performance and Durability Assessment*, Elsevier, 2004.

[56] D. C. Jordan, T. J. Silverman, B. Sekulic, and S. R. Kurtz. PV degradation curves: Non-linearities and failure modes, *Progress in Photovoltaics: Research and Applications*, 25 (7), 2017, 583–591.

[57] I. Kaaya, S. Lindig, K.-A. Weiss, A. Virtuani, O. M. Sidrach de Cardona, and D. Moser. Photovoltaic lifetime forecast model based on degradation patterns, *Progress in Photovoltaics: Research and Applications*, 2020.

[58] I. Kaaya, M. Koehl, A. P. Mehilli, S. de Cardona Mariano, and K. A. Weiss. Modeling outdoor service lifetime prediction of PV Modules: Effects of combined climatic stressors on PV Module power degradation, *IEEE Journal of Photovoltaics*, 9, 2019, 1105–1112.

[59] J. Ascencio-Vasquez, I. Kaaya, K. Brecl, K. A. Weiss, and M. Topıc. global climate data processing and mapping of degradation mechanisms and degradation rates of PV modules, *Energies*, 12 (24), 4749, Dec 2019.

[60] F. K. A. Nyarko, G. Takyi, and E. H. Amalu Robust crystalline silicon photovoltaic module (c-Si PVM) for the tropical climate: Future facing the technology, *Scientific African*, 8, 2020, e00359, doi:10.1016/j.sciaf.2020.e00359.

[61] S. Lindig, I. Kaaya, K.-A. Weis, D. Moser, and M. Topic. Review of statistical and analytical degradation models for photovoltaic modules and systems as well as related improvements, *IEEE Journal of Photovoltaics*, 8, 2018, 1773–1786.

[63] A. Phinikarides, N. Kindyni, G. Makrides, and G. E. Georghiou. Review of photovoltaic degradation rate methodologies, *Renewable and Sustainable Energy Reviews*, 40, 2014, 143–152, doi:10.1016/j.rser.2014.07.155.

[64] P. Bhola and S. Bhardwaj. Clustering-based computation of degradation rate for photovoltaic systems, *Journal of Renewable and Sustainable Energy*, 11, 2019, 014701, doi:10.1063/1.5042688.

[65] J. M. Howard, E. M. Tennyson, B. R. A. Neves and M. S. Leite. Machine learning for perovskites' reap-rest-recovery cycle, Joule, 3, 2019, 325–337, doi:10.1016/ j.joule.2018.11.010.

[66] A. Bala Subramaniyan, R. Pan, J. Kuitche, and G. Tamizh Mani. Quantification of environmental effects on PV module degradation: A physics-based data-driven modeling method, *IEEE Journal of Photovoltaics*, 8 (5), 2018 September, 1289–1296.

[67] I. Kaaya, J. Ascencio-Vasquez, K.-A. Weiss, and M. Topıc. Assessment of uncertainties and variations in PV modules degradation rates and lifetime predictions using physical models, [To be published], 2020.

[68] M. Koehl, S. Hoffmann, and S. Wiesmeier. Evaluation of damp-heat testing of photovoltaic modules, *Progress in Photovoltaics: Research and Applications*, 25 (2), 2017, 175–183.

[69] J. Zhu, M. Koehl, S. Hoffmann, K. A. Berger, S. Zamini, I. Bennett, E. Gerritsen, P. Malbranche, P. Pugliatti, A. D. Stefano, F. Aleo, D. Bertani, F. Paletta, F. Roca, G. Graditi, M. Pellegrino, O. Zubillaga, F. J. C. Iranzo, A. Pozza, T. Sample, and R. Gottschalg. Changes of solar cell parameters during damp-heat exposure, *Progress in Photovoltaics: Research and Applications*, 24 (10), 2016, 1346–1358.

[70] R. Pan, J. Kuitche and G. Tamizhmani. (2011, January). Degradation analysis of solar photovoltaic modules: Influence of environmental factor. In 2011 Proceedings – Annual Reliability and Maintainability Symposium, 1–5. ISSN: 0149-144X, 0149-144X

[71] B. Braisaz, C. Duchayne, M. Van Iseghem, and K. Radouane. (2014, September). PV aging model applied to several meteorological conditions. In 29th PVSEC Proceedings, Amsterdam, 2303–2309.

[72] L. A. Escobar and W. Q. Meeker. (2007, August). A Review of Accelerated Test Models. arXiv:0708.0369 [stat]. arXiv: 0708.0369.

[73] Copernicus climate change service (C3S) ERA5: Fifth generation of ECMWF atmospheric reanalyses of the global climate. Copernicus Climate Change Service Climate Data Store (CDS) 2017. Available online: https://cds.climate.copernicus.eu/cdsapp#!/home (accessed on 13 October 2019).

[74] J. Brownlee Introduction to Time Series Forecasting with Python: How to Prepare Data and Develop Models to Predict the Future. Machine Learning Mastery; 2017.

[75] DIN EN ISO 14040: 2009–11: Umweltmanagement – Ökobilanz – Grundsätze und Rahmenbedingungen.

[76] DIN EN ISO 14044: 2006: Umweltmanagement – Ökobilanz – Anforderungen und Anleitungen.

[77] JRC European Commission. Recommendations for Life Cycle Impact Assessment in the European Context. ILCD Handbook; JRC European Commission: Ispra, Italy, 2011; ISBN 978-92-79-17451-3.

[78] Product Environmental Footprint Category Rules (PEFCR); Photovoltaic modules used in photovoltaic power systems for electricity generation. Version: 1.1; February 2019; https://ec.europa.eu/environment/eussd/smgp/pdf/PEFCR_PV_electricity_v1.1.pdf

[79] S. Pinto Bautista. Influence of PV waste management on the environmental footprint of electricity production from photovoltaic systems, 2019, Master thesis.

[80] S. Herceg, S. Pinto Bautista, and K.-A. Weiß. Influence of waste management on the environmental footprint of electricity produced by photovoltaic systems, *Energies*, 13, 2020, 2146.

[81] S. Weckend, A. Wade, and G. Heath. 2016. End-of-Life Management: Solar Photovoltaic Panels. Hg. v. IRENA and IEA-PVPS (ISBN 978-92-95111-98-1).

[82] R. Kemna. 2011. Methodology for Ecodesign-related Products – Project Report. MEErP 2011. Hg. v. COWI Belgium SPRL – in association with – Van Holsteijn en Kemna B.V. (VHK). European Commission.

[83] American National Standards Institute (2019): NSF/ANSI 457–2019. Sustainability Leadership Standard for Photovoltaic Modules and Photovoltaic Inverters. Hg. v. NSF International.

[84] ASTME313, *Standard Practise for Calculating Yellowness and Whiteness Indices from Instrumentally Measured Color Coordinates*, Pennsylvania, USA, 2000.

[85] C. Peike, W. Phondongnok, T. Kaltenbach, K.-A. Weiss, and M. Koehl. Non-destructive determination of the cross-linking degree of EVA by Raman Spectroscopy, *Open Journal of Renewable Energy and Sustainable Development*, 1 (1), 2014, 14–21.

[86] K. Wambach. Life cycle inventory of current photovoltaic module recycling processes in Europe; IEA PVPS Task12, Subtask 2, LCA Report IEA-PVPS T12-12:2017; National Renewable Energy Lab. (NREL): Golden, CO, USA, 2017.

Index

https://doi.org/10.1515/9783110685558-017